SOCIAL MEDIA IN DISASTER RESPONSE

Social Media in Disaster Response focuses on how emerging social web tools provide researchers and practitioners with new opportunities to address disaster communication and information design for participatory cultures. Both groups, however, currently lack research toolkits for tracing participant networks across systems; there is little understanding of how to design not just for individual social web sites, but how to design across multiple systems. Given the volatile political and ecological climate we are currently living in, the practicality of understanding how people communicate during disasters is important both for those researching solutions and for those putting that research into practice.

This situation is addressed by presenting the results of a large-scale sociotechnical usability study on crisis communication in the vernacular related to recent natural and human-made crisis; this is an analysis of the way social web applications are transformed, by participants, into a critical information infrastructure in moments of crisis. *Social Media in Disaster Response* provides researchers with methods, tools, and examples for researching and analyzing these communication systems while providing practitioners with design methods and information about these participatory communities to assist them in influencing the design and structure of these communication systems.

Liza Potts is an assistant professor in the Department of Writing, Rhetoric, and American Cultures at Michigan State University. She is a senior researcher at Writing in Digital Environments Research Center, the director of user experience at MATRIX, and a collaborator at Creativity Exploratory—a practice-based addition to the College of Arts and Letters curriculum. Her research interests include technologically mediated communication, experience architecture, and participatory culture. Potts is the chair of the Association for Computer Machinery's Special Interest Group on Design of Communication (ACM: SIGDOC) and the co-editor of *Communication Design Quarterly Review*. She has worked for Microsoft, consultancies, and start-ups as a director, user experience architect, and program manager.

ATTW Book Series in Technical and Professional Communication

Jo Mackiewicz, Series Editor

Social Media in Disaster Response: How Experience Architects Can Build for Participation
Liza Potts

For additional information on this series please visit http://www.attw.org/ publications/book-series, and for information on other Routledge titles visit www.routledge.com

SOCIAL MEDIA IN DISASTER RESPONSE

How Experience Architects Can Build for Participation

By Liza Potts

Routledge
Taylor & Francis Group

NEW YORK AND LONDON

First published 2014
by Routledge
605 Third Avenue, New York, NY 10017

and by Routledge
4 Park Square, Milton Park, Abingdon, Oxon OX14 4RN

Routledge is an imprint of the Taylor & Francis Group, an informa business

Library of Congress Cataloging-in-Publication Data

Potts, Liza.
Social media in disaster response : how experience architects can build for
    participation / by Liza Potts.
       pages cm. — (ATTW/Routledge book series in technical and
professional communication)
  1. Emergency management.   2. Social media.   3. Online social
networks.   I. Title.
  HV551.2.P68 2014
  363.34′802854678—dc23
  2013028029

ISBN: 978-0-415-81742-4 (hbk)
ISBN: 978-0-415-81741-7 (pbk)
ISBN: 978-0-203-36690-5 (ebk)

Typeset in Minion
by Apex CoVantage, LLC

To the participants of the many disasters I have traced, all of whom have reminded me what a gift the social web—both online and off-line—can be in our darkest times of need. And to my girls, Zoë, Katie, and Jayne, who remind me daily of what a gift life can be.

CONTENTS

FIGURES

SERIES EDITOR FOREWORD

With this book—Liza Potts's *Social Media in Disaster Response: How Experience Architects Can Build for Participation*—we begin what will be a long line of research-driven, useful, usable, and interesting texts that will comprise the ATTW Book Series in Technical and Professional Communication. As some readers will remember, Potts's book and the books that will follow it are not the first that ATTW has published. In the 1980s and 1990s, ATTW published Carolyn Rude's *Teaching Technical Editing*, John Harris's *Teaching Technical Writing*, and Charles Sides's *Technical Communication Frontiers: Essays in Theory*, among others. But as readers of *Technical Communication Quarterly* (ATTW's scholarly journal) know, the skills, knowledge, and practices that fall within the scope of technical and professional communication (TPC) today are far more diverse than they were years ago when ATTW published those books. Concomitantly, the diversity of ATTW members' interests has grown as well. It is fitting then—indeed, it is critical—that as one of our field's most important professional organizations, ATTW articulate questions about current issues and make headway in answering those questions. With this book series—developed through the effort and tenacity of Stuart Selber and other ATTW Executive Committee members—ATTW underscores its commitment to TPC professionals of all sorts.

The goal of the ATTW Book Series in TPC is simple to state, but it is difficult to achieve. As I noted previously, the goal is to publish research-driven, usable, useful, and interesting books. I envision print books that are left close-at-hand on desks and that become dog-eared with use rather than being sentenced to line up on the shelf. I envision e-books covered in highlighting and enriched with notes—notes that spark conversation as readers share them with others. In other words, the goal is not to create another outlet for publishing academic treatises that have value for and hold the interest of a few specialists and no one else.

Developing research-driven, useful, usable, and interesting books means working with authors to make abstract theory and complicated research comprehensible and to make explicit the possible implications of research findings for TPC pedagogy and practice. Thus, my aim as the first editor of this series is not only to help authors develop their ideas but also to help them identify and delineate the ways that those in industry and academia might put their ideas to use.

Our series will freshly address topics that have been and continue to be interesting to TPC professionals in industry and academia, including plain language, program assessment, publication management, visual communication, and editing. But the series will also explore emerging issues such as professional discourses in science and medicine, the relationship between marketing and technical communication, and sustainability in technical communication. Such topics reflect the indeterminate boundaries of our field, as well as the broad territory it spans.

So too do the field's theoretical foundations and its research methods reflect its network of diverse but interrelated and intertwined knowledge and practices. We in TPC come from undergraduate and graduate programs in rhetoric, linguistics, education, and now more commonly technical and professional communication. The theories we draw on range from activity theory to politeness theory to social development theory. We employ methods such as corpus, discourse, and rhetorical analysis. We use experimental and ethnographic methods and everything in between. Such diversity can pose a challenge when we need to explain our field of study to those who are unfamiliar with it, but most of us I think would agree that our field's diversity is far more of a blessing than a curse: we can become—if we so choose—something like renaissance people in our work.

Starting with Liza Potts's *Social Media in Disaster Response,* our ATTW Book Series in TPC takes advantage of and supports the breadth and depth of our many interests. The series publishes two lines of books: practice and research. Books in the practice line will focus on teaching and learning in industry and in communication classrooms. Practice-line books are more likely to contain chapter-ending exercises, prompts suitable for students in upper-division or graduate courses and for those in industry who are either learning on their own or participating in company-sponsored communication training. Also, the practice-line books will contain robust supplementary material such as online examples, templates, and practice assignments. Although authors of research-line books will discuss and develop theory, they will, like the authors of practice-line texts, delineate and describe the implications of their work. Potts's research-line *Social Media in Disaster Response* exemplifies this commitment to explicitly connecting research findings to their potential applications. Potts concludes chapters 3, 4, and 5 with sections on practical solutions.

That Potts's *Social Media in Disaster Response* launches the series is also fitting in that it analyses a type of technical communication deserving more analysis than it has heretofore received—communication that laypeople create, revise, and repurpose, in this case in crisis situations. In addition, Potts's book examines the ways in which laypeople use emergent social media to gather and share information. That is to say, *Social Media in Disaster Response* represents the needed and timely books that this series will develop.

As the series editor, I look forward to working with authors to develop books like Potts's *Social Media in Disaster Response*—readable, interesting books of use to practitioners and academics in TPC and related fields.

Jo Mackiewicz
Editor, ATTW Book Series in
Technical and Professional Communication
6 March 2013

ACKNOWLEDGMENTS

Disasters are chaotic, unsettling, and utterly disturbing. For nearly 10 years, I have documented over 50 events—acts of terrorism, natural disasters, school shootings, attempted bombings, and more. What has kept me going through this project is knowing that there is so much good in our world. Watching everyday people rush to help each other, witnessing the bravery of first responders, and hearing stories of real heroism has been nothing short of amazing. I have been there with these participants as much as I could be, trying to help track down information and validate sources. These tasks have not been easy, but what I have seen has renewed my hope in our humanity and the social web.

As I prepared this book, plenty of people, both online and off-line, contacted me about disaster reports, interesting websites, or hashtags I should follow. Doctor Disaster thanks you all. During my time working on disasters, I have met some amazing volunteers and social web participants. Special thanks in particular to Joyce Seitzinger, who brought her research skills and firsthand experiences to the #eqnz event, and to Dina Mehta, for her amazing work during the Mumbai attacks. Throughout this project, I have had the honor of working with many awesome colleagues and students. Thank you to everyone who read and commented on the drafts of this book and the articles leading up to it: Cheryl Geisler, Kathie Gossett, Bill Hart-Davidson, Loel Kim, Clay Spinuzzi, Kirk St. Amant, Jason Swarts, and my anonymous peer reviewers. Thank you to my awesome graduate students, who experienced many versions of the discussions in this book: Dave Jones, Angela Harrison Eng, Beth Keller, Tiffani Bryant, and Robin Ormiston. Thank you to March Rosenbluth and Beth Keller for helping me finalize the ANT (actor-network theory) diagrams. For the final push of editing, thank you to Abbi Lynch, Katie Dobruse, and Chelsea Moats. Thank you to the series editor, Jo Mackiewicz, and to Linda Bathgate at Routledge, for their support and guidance. Finally, thank you to my mentors at Michigan State: Jeff Grabill, Laura Julier, and Dànielle DeVoss. This book was generously supported by Michigan State University's Humanities and Arts Research Program grant.

This book got its spark as a dissertation back in the dark ages of the mid-2000s when I needed a glossary to define "social software" and "blog." So much has changed since then; we've seen new tools, new spaces, and, unfortunately, many more disasters. Many thanks to Bill Hart-Davidson, June Deery, Cheryl Geisler, Jim Zappen, and Katherine Isbister for their support at Rensselaer Polytechnic Institute. Thank you, June, for ensuring that I kept the ideas of culture, media, and participation foremost in my work—your influence continues to bring a level of humanity to this project that would not be here without your guidance. Many thanks to Bill Hart-Davidson, an amazing mentor and friend whose teachings, insights, and discussions had a major impact on my dissertation, this book, and the scholar I am today.

On a personal note, I would not be half the researcher or architect I am were it not for my partner, spouse, and co-conspirator: Steve Potts (SEGA). To my Potts girls, thank you for helping me realize how wonderful life can be. And thank you to my mother, Carmen Kaplan, who kept me fed, watered, and entertained throughout.

1

EXPERIENCE, DISASTER, AND THE SOCIAL WEB

From 26 to 29 November 2008, the Mumbai attacks in India caused countless social web participants to mobilize, locating data, verifying information, and distributing knowledge across the globe. This terrorist event resulted in 166 civilian deaths and at least 304 injuries. Across the city, terrorists hit 12 locations (Chief Investigating Officer, Government of India 2009). From explosions and gunfire at the Taj Mahal Palace hotel to the atrocities committed at the Nariman House, they affected numerous transportation systems and tourism sites. Locations as diverse as cafés, hotels, hospitals, and railway stations were attacked. Throughout this multiday event, many volunteers online and reporters in the mainstream media took on the challenging knowledge work of locating and validating information from eyewitnesses.

Mumbai was a major tipping point for online participation during times of disaster. From the use of Google Docs and Twitter to the uploading of numerous images to Flickr, this event began a new era for the social web—an era that systems developers can no longer ignore. The industry and the field must move past thinking within their single-serving interfaces, systems, documents, and silos if they are to meet the needs of new social knowledge workers. While these platforms were not built to move information across systems, participants invented connections in order to find and share information related to the attacks. And while participants did their best to bridge these systems, they encountered many obstacles and challenges as they attempted to exchange information about the attacks.

In this book, I define social web ecosystems, explain methods for researching the participatory use of these systems, and discuss the need for an interdisciplinary team led by experience architects to build them. Experience architects look across ecosystems, building for experiences that move across spaces, systems, and networks. I demonstrate how we can trace the ways everyday people deploy social web technologies to share and exchange content. In the case studies in this book, I specifically examine disaster situations. However, I also explain how we can generalize these experiences to other kinds of use and experience within digital spaces. I explain why humanists and technologists are especially well placed to lead this tracing and building—primarily because of our understandings of culture, use, and context. We can coordinate—if not guide—the activities of design and development teams. Lastly, this book discusses in detail the ways we can bring about better architectures for everyday social web situations. I make these moves by explicitly resituating this work as experience architecture, pointing to the need for understanding the experiences of participants who leverage these technologies.

Such an examination is, in itself, an interdisciplinary task; any analysis of the building tools for the social web must take into account the many systems that mediate communication among people, organizations, and technologies. By discussing the need for shared perspectives across humanities and sciences, industry and academia, this book calls for the kinds of research and practice that can move us beyond traditional practices and into the new paradigm of the experience architect.

Architecting Mediated Systems

In this book, I set out to define, illustrate, and encourage the architecting and researching of mediated systems for the social web—systems built for holistic experiences that span multiple technologies, people, and organizations. They are systems that help mediate communication by interacting, structuring, and architecting the flow of information among actors. We can build such systems if we understand them as participatory ecosystems that must allow for flexibility and responsiveness.

When discussing the process of architecting mediated systems, the map of the London Underground is a useful example. Originally designed by Harry Beck in 1933, this popular illustration of the London Underground (i.e., the London subway system) has been updated and is currently in use by Transport for London (2012). This map is an excellent illustration of the ways networks can be abstracted for those who use them. The map represents an imagined London—the points on the map do not directly correlate with the physical geography aboveground. This imagined London, however, provides commuters with a tangible artifact that they can identify with to connect the concept of physical space with their personal transportation needs. In discussing the map's significance, travel author Bill Bryson (2001) notes how Beck "created an entirely new, imaginary London that has very little to do with the disorderly geography of the city above" (41). Although it could potentially mislead travelers since it does not match geographical reality, Beck's work better communicates how to navigate the Underground. This map creates a world that brings a sense of order that better serves the needs of its audience. That is, the map creates a sense of order in symmetry.

Much like these maps, social web systems often represent a disconnect between what practitioners design for and what individuals actually need in these systems.[1] Our job as researchers and practitioners is to help bring order out of chaos in these imagined spaces. Any viable solution must push beyond the current user-centered design paradigms if we are to architect systems that provide solutions.

Experience Architecture

Although many of the case studies that I examine in this book focus on experience architecture issues, the problems associated with using these technologies are not purely issues of user interface, interaction design, and information design. In fact, broader issues for experience architects emerge in social web experiences. Other problems relate to systems that lie beneath the surface—systems that dictate data structures, permit reuse, or allow for sharing across systems. For the sake of sanity, throughout this book, I refer primarily to "experience architecture," although I also use terms such as "information design," "user experience," "design," "communication design," and "technical communication." Unless I am making an explicit distinction, I use these terms synonymously to discuss the architecture of the systems both above and below the surface (i.e., architecture of interactions, visuals, content, structure, and policy).

This concept of experience architecture is not particularly new. What is new is the idea that we need to work in teams with diverse backgrounds in writing, design, development, and information systems to build them. This work is not simply designing visuals or coding features; it focuses on architecting the end-to-end experiences of the participants who will be using these systems. Thinking outside of a single use where people sit in one program all day long, experience architects look at how an individual component is part of a larger ecosystem. These ecosystems include multiple technologies, devices, websites, organizations, people, and events. Social web tool use—from including links to related material in retweets on Twitter to tagging people in Facebook photos to posting video responses in blog comments—clearly shows that participants are linking together multiple systems to forage for information and distribute knowledge. For humanists to have a stake in advancing these systems, they must be prepared to address issues of interface design, information architecture, findability, usability, and other experience concerns.

In my own work building digital products and services in industry and academia, I have hired many humanists to lead projects, act as information architects, and lead content development groups. We experience architects are already performing these roles on teams, buoyed by our humanities and social science backgrounds that have prepared us for understanding culture, use, and experience. Examining the case studies in this book moves the field and the industry one step further in pursuit of this goal and supports the existing work of information architects, information designers, and software developers.

Interdisciplinary Work

Architecting systems for the mediations among people, technologies, and groups demands perspectives from various backgrounds, educations, and disciplines. Such perspectives are even more vital as these systems permeate our social, business, and research lives. As mentioned earlier, this work lies both within the scope of what we are already doing in more traditional technical communication and across the broader groups of developers, designers, and architects. Working in teams to research, prototype, test, and deploy, these groups are in desperate need of more information on how to improve their products and services. Humanists are best located at the center of these movements because of their ability to understand users, manage projects, and work with diverse team members. We, as scholars, practitioners, and teachers, are trained to focus on persuasion, context, audience, and reception. Such training makes humanists uniquely situated for these kinds of interactions, and we must move from the sidelines to the center of these discussions.

Drawing on research found at the intersection of multiple disciplines—specifically technical and professional communication, participatory culture studies, and science and technology studies—we can trace how everyday people deploy technologies to communicate with each other. Tracing communication illustrates how architects can lead the analysis and design for both the social and technological infrastructures of communication by framing new practices, creating new tools, and enabling richer and more valuable interactions. Experience architects can trace these activities by championing participant-centered research methods and by becoming active participants in the communities for which they design. In doing so, they can practice the kind of holistic, context-rich experience architectures necessary to create systems that allow information to flow between people and technologies.

To build systems that serve the participants' needs, experience architects need to hear from the users, participants, researchers, designers, developers, business analysts, and various other stakeholders. They also need to participate within these spaces as participants themselves. Such interaction is essential because stakeholders play a role in building the participant experience; their input is essential for ensuring appropriate, necessary interactions between participants and systems. Research teams need to develop shared perspectives that extend across humanities and sciences, industry and academia, and design and engineering. And experience architects, who straddle these diverse disciplines, are in the perfect position to organize—if not lead—the activities of such teams.

Moving across Boundaries

By integrating practices from both academia and industry that are active in mediated systems architecture, developers and designers move toward creating flexible systems that support people-powered solutions. In positioning our work at the intersection of systems design, technical communication, and digital culture, experience architects provide a newer and richer method for conceptualizing and creating mediated experiences for a range of users. Researchers and designers must embed themselves within social web activities in particular.

By embedding themselves within these scenarios, experience architects will understand the frustrations of participants who must negotiate often-broken systems and manipulate difficult-to-use tools in order to communicate as effectively as possible in a high-stakes situation. Going beyond observation, researchers and content developers must become these participants to understand how these mediations are succeeding and failing. By following hashtags, retweeting important news, adding tags to images, tracking missing persons through news reports, commenting on blogs, and witnessing how participants stretch social web systems in ways the authors never intended, experience architects gain an understanding that simple observation alone would never provide. Such insights lead to architecting the kinds of solutions that allow for quicker adoption of these tools when we need them most: the moments after a disaster occurs.

When using the term "architecture" in this book, I refer to the entire participant experience. I also refer to the information architecture that supports the data structures and naming conventions within networks (Morville and Rosenfeld 2006), the interaction design that structures the users' flow of activity across these systems (Cooper, Reimann, and Cronin 2007), and the content strategy that organizes and presents relevant information for participants (Halvorson and Rach 2012; Redish 2007). All of these experience architecture issues are interdependent; no one problem can or should be isolated from the others. Many academic programs with this worldview have evolved so that industry will highly value their graduates.

For the theory and the practice of experience architecture to progress, processes and structures must be in place to support building for participatory exchange. For architects to help build such systems, they must reevaluate current software design paradigms with an eye toward building participatory experiences. In essence, we must rethink both the concepts of how people interact with systems and the perceptions of the designer's place within these activities (Potts 2009b; Spinuzzi 2003).

The Social Web

Before I outline how to architect systems for the social web, we must examine how communication design practices need to shift to support participatory culture. In this section, I discuss these terms, define them as they relate to academia and industry practice, and explain why understanding these definitions is critical for project teams to work productively across disciplines. Using instances of disaster as a backdrop, I define these concepts in terms of a relevant, timely need.

We have an ethical, political, and philosophical obligation to equip current and future architects with methods that inform the research and design of systems that can help support the social web. This responsibility exists both for supporting everyday use of social web systems and for the high-stakes, high-pressure experiences that occur within disaster and crisis situations (Potts 2009a). This obligation exists for those who architect or have influence on the design of these systems, instruct others on how to build them, or research their development and social implications. To quote Sullivan (1990), we must "see ourselves as doing more than teaching a set of skills." Our obligation "places ethical and political responsibility upon us" to answer this call with appropriate methods to address these situations (384). We must cease building antisocial software that works to instruct users on what they can and cannot do in these spaces in favor of building systems that are socially flexible, allowing participants to flourish.

Defining the Social

In considering this evolution, we need to look at the social web, which industry and academia have referred to as "social networking systems," "social software," and "social media," depending on their perspectives and intentions. "Social networking systems" refers to the actual systems that allow people to form, join, and participate in networks such as Facebook. A term coined in 1987 (Drexler 1991) and made popular in the 2000s, "social software" refers to these tools as artifacts, noting their interfaces, interactions, and structures. "Social media" sometimes also describes these systems, but the term is also aligned with marketing campaigns and the rise of these tools as a media outlet. Various media outlets also refer to these technologies (somewhat tiringly) as "Web 2.0," although the popular press more often uses the term "social media."

In this book, I use these terms interchangeably when discussing internet-based networking systems that provide spaces for people to build communities, however temporary they might be. More specifically, I refer to these systems in general as the "social web," sometimes specifically as "social web tools." At first glance, this definition of social web seems broad. Yet, similar usage patterns of social sharing and knowledge exchange are apparent across a multitude of these sites. On instant messenger chat, participants can quickly exchange details one-to-one. On media-sharing sites such as YouTube, Vimeo, and Flickr, participants can post video feeds and photos of disaster tours and activism. On social networking sites such as Facebook, people can create groups based on disaster locations and events, as they did in the wake of the Virginia Tech massacre. On social blogging systems such as Tumblr, LiveJournal, WordPress, and Blogger, participants can collect information and write first-person accounts of the disasters.

Using information-networking tools such as Twitter, knowledge workers around the globe can quickly exchange messages. On social knowledge repositories such as Wikinews and *Wikipedia,* participants can collect information and validate the content with what is appearing on other trusted sites. Lastly, but by no means exhaustively, collaborative writing spaces such as Google Docs allow participants to structure information and share writing tasks across networks.

Rather than studying or architecting for single, one-to-one actions between users and individual tools, I look at these participatory experiences holistically, taking into account actions made to share knowledge across multiple technologies. As Geisler (2011) states, "Twitter tools and Web-scale texts, on the other hand, provide more opportunities for readers and writers to become immersed in a virtual experience beyond what is familiar to them—not, however, to escape their embodied reality but to extend it" (253). Such immersion can lead to the kinds of products and services that are open to participation and engagement.

Within such context, architects must be active and participate in systems. To achieve this goal, architects must move away from the static notion of early web implementations to the interactive, social technologies that users demand. Building sites that allow participants to tag content with metadata, label information with relevant details, and reply to participants and developers is critical for such participation to take place. To recognize participants as stakeholders in the growth of these sites is key to forming spaces in which such activities can flow more freely. Blurring the lines between designer and participant, producer and audience, and writer and reader should be seen as opportunities to create technologies and policies that empower this level of engagement on all sides.

So much of our early research on social web tools looked at issues of community, identity, access, and refinement of definitions (Baym 2000; Doheny-Farina 1996; Donath 1999; Ess 1996; Gauntlett 2000; Herring 1993; S. Jones 1999; Rheingold 1993; Turkle 1995; Wellman 1997). In many of these articles, chapters, and reports, researchers focused on technologies such as email, chat, and online bulletin board systems because of the time period in which they were written. Research has shifted from looking solely at textual issues to looking at issues of interface, interaction, and architecture. More recent work has, in turn, looked at interaction design, media convergence, participation, and knowledge sharing, whether for work, entertainment, or other pursuits (Baym 2010; boyd 2007; Bruns 2008; Halavais 2009; Jenkins 2006; Potts 2010; Stolley 2009; van Dijck 2009). With 81% of adults in the United States now online (Pew Internet and American Life Project 2012b), researchers are trying to assess information processing and participatory culture. The fast-paced evolution of spaces has increased the opportunity for further scholarly work and for further innovative design work.

In 1997, Jorn Barger created the first blog ever (Wortham 2007). In August 2011, 180 million US viewers watched video content on sites such as YouTube and Hulu (comScore 2011). In 2012, photo-sharing website Flickr featured more than 7 billion photos (Leung 2012). Today, the English-language version of *Wikipedia* contains over 4.1 million articles (*Wikipedia* 2012). Facebook has surpassed 1 billion monthly active users (Facebook 2012). When I began the research for this book, Twitter did not exist, but in 2012, Twitter had surpassed 200 million active users (Twitter for Business 2012). All of these statistics clearly indicate that the social web has become a hive of communicative activity.

Shifting from Use to Participation

Despite the time that has passed since social web tools were created, a statement made by an early blogger in regard to the web is still relevant to describe how today's participants feel about social networking: the "mother lode of personal expression—the one place in our lives that we (as people lucky enough to have access) can say whatever we want about anything we want" (Powazek 2000). And, as I describe it in this book, the social web has become an important space for saying what we need or want to say and for saying it when it is most relevant.

The terms "user" and "participant" sound similar, but they have very important differences that are critical to understanding this shift in culture and use. Within the context of this book, I employ the term "users" when discussing the use of a system's technology (e.g., single-task systems such as word processors and spreadsheets). I use the term "participant" to emphasize participatory and community-oriented users who leverage their activities as points of mediation (e.g., writing articles in *Wikipedia* and posting to Twitter while using hashtags). This distinction is important. While referring to people as users is easy, doing so undermines the notion of how centrally important participation has become in systems. As the case studies in this book reveal, during times of disaster, users created their own spaces in ways the designers never intended. In so doing, they employed photo-sharing sites to locate affected locations, relied on news sites for distributing missing-persons information, and used status-update sites as catalysts for distributing community knowledge. Given the enthusiasm individuals displayed in using mediated systems to share information in times of crisis, participants seem to now be partners in the co-creation of these communication tools and the social conventions within them.

Yet, not all of these communities emerge in the same way. For example, longstanding internet communities such as those found in professional circles on LinkedIn, fan cultures on LiveJournal, or the wide range of groups on Facebook pages often involve long-term connections. In times of crisis, however, communities tend to erupt around a given disaster and disperse quickly afterward. Existing sometimes over a series of days or weeks rather than months or years, these microcommunities form to help each other work through a disaster, share much-needed information, or organize other types of responses during and after a crisis. Any shortened existence of these communities stems not from a lack of tools present in current technologies but from the fact that disasters lead to questions that are either answered within days after the event ("Is my missing friend okay?") or that require answers that cannot be uncovered quickly through these tools ("How long will it take to rebuild New Orleans?"). For the most part, this shortened existence is why I avoid the term "community," instead favoring terms such as "group" or other specific words regarding a particular case study. While such entities appear more like temporary clusters than long-standing communities, these groups of participants exhibit the similar traits in data gathering, information coordination, and knowledge sharing that more stable online communities possess.

Disaster, Communication, and the Social Web

I chose to examine disaster case studies in this book for specific reasons. Like any researcher, I needed data sets with strong signals that could provide enough material to show significance and that would examine a compelling problem. That's the simple reason. Actually digging in to these case studies is far more complicated. In the chaos and panic of a disaster, persons using social web tools to communicate and validate information are worthy of study because they have been able to use the available tools in new ways to make connections. In a limited time, they must jump into communities, moderate participants, locate information, validate it, and redistribute it to other participants. Participants are constructing new paradigms for participatory design, encouraging a shift toward active exchange rather than passive use. Observing and participating within systems encourages information architects to learn how to better construct them for everyday communication. These high-pressure, high-stakes cases provide a view into holistic experience, where participants use multiple systems while seeking out information and other participants. Examining these events by looking across multiple sites such as Flickr, Craigslist, LiveJournal, Facebook, Twitter, blogs, and wikis allows experience architects to experience activities in ways similar to how participants experience them.

In my work, I have spent nearly a decade studying the use of social web tools in the wake of mass shootings, hurricanes, earthquakes, acts of terrorism, and other disasters. Much of this work has been difficult, and it has often been disturbing. It has by no means been easy to witness these events as they unfold, trace the missing, and participate with people desperate to communicate. I have seen photos, watched videos, traced streams, and read posts that have shaken me to my core. However, the connections I have made and the solutions I have offered have been completely and utterly worth that cost. And certainly, I have by no means been through anything close to what the victims, friends, and relatives of those affected by disaster have been through. They have my utmost respect and awe for what they have witnessed, what they have accomplished through the work they do, and how they have survived.

The case studies that I analyze draw upon various disasters, both natural and human-made, that have taken place in the early twenty-first century. This selection provides researchers with a relatively recent look at how understanding information flow and systems can facilitate communication. The power of these cases, moreover, lies not only in their ability to motivate designers to attend to them because of their seriousness. Rather, their power is deeply connected to the patterns of use that experience architects can derive from them. As a result, the analysis of these cases can help researchers and practitioners rethink their role in relation to the social web and its uses outside of these specific disaster cases. Generalizing these examples, the content in these chapters also points to practical solutions for the concerns raised in the case studies.

That said, this research is generalizable to social web architecture and the software design process. Cobbling together these networks and various specialty sites, the moderators of these networks—the collectors of information—attempt to pull data out of the confusing array of silos in which they find the data. They work to validate the data they find, retooling this information for the community as shared knowledge. Not surprisingly, participants in the social web, especially individuals engaged in networks that emerge in response to a crisis, are always on the move. These persons move among many different sites, across dispersed pieces of information, and through various and often disconnected resources for managing, organizing, linking, and sharing that information; these individuals move quickly and effectively to make the information they are employing more useful to those who need it. Thus, disaster victims frequently turn to the internet during emergencies, particularly when phone communications are disrupted (Putnam 2002). Victims use the web to reach out to loved ones, find friends and family, or simply find help from the Red Cross or other emergency relief organizations. In essence, they have adapted and expanded the older practice of using media to "reach out and touch someone." And every day, people across the social web do this kind of activity, whether it is updating their status on Facebook, adding a new photo to Flickr, or tweeting on Twitter.

Many disaster researchers have focused their efforts on analyzing the content and actions of television and radio media during crises (Blakemore and Longhorn 2001). Others have turned their research toward the actions of citizens who respond to these disasters (Carey 2003; Collins 2004; Rappoport and Alleman 2003; Vengerfeldt 2003). In contrast, the research I present points to a new and always-evolving space in which the activity of participants can change from one event to another, from one task to another, or even from one moment to the next. Average disaster victims cannot broadcast via television and radio to the extent they can via the social web; this distinction highlights how radically different the research—and the findings—in this book are from past research on this topic. Internet-based communication tools can appear, change, or even disappear rapidly. To better understand the ways social web participants use systems to respond to such emergencies, experience architects must develop new lenses with which to view such phenomena.

Experience architects have the requisite knowledge and skills needed to map these networks, learn what information is important to social web participants, uncover how they spread that information, and then drive the design decisions that can make social software much more useable, especially when it is used in response to disasters (Potts 2008, 2010). Our responsibilities as architects thus spread beyond the hallways of our respective academic institutions and industry offices. The practical focus of our work as participant-centered researchers and practitioners empowers us to improve experiences so that in extreme moments families can more easily locate their loved ones and in everyday moments people can communicate in more effective ways.

If readers glean only one idea from this book, I hope it will be this: in times of disaster and crisis, people tend to gravitate toward the systems and networks that are most relevant and familiar to them. They shun the specialized sites set up by aid workers for the comfort of already-established systems they know well from prior experience. Their participation in these sites in times of crisis is, of course, typically outside of their everyday communicative activities. What happens is a metamorphosis of form that produces a new and more exigent function. Familiar sites for sharing photos of weddings become locations for sharing breaking news, social networks meant to connect college students become locations to post the names of the deceased, and collaborative writing tools become locations to transcribe hospital faxes to confirm the injured. Participants are using social web tools in ways the designers of such systems had neither anticipated nor considered. As a result, the vast majority of today's technologies are woefully ill-equipped for crisis situations; they narrowly support everyday communications. In this book, I analyze examples of social web tool use in crisis situations in order to identify patterns and to discuss how experience architects might improve the structure and use of such technologies for participation.

Disaster Cases

The following disasters, some natural and others human-made, guide the cases that I discuss. I selected each case to highlight the theme of a chapter: data, information, or knowledge. In these examples, disasters affected populations throughout various international locations. These cases serve as watershed moments for the kinds of participation that necessitate a rethinking of research and design methods and tools. While I mention other disasters throughout this book, these three cases receive strong attention.

Hurricane Katrina

In the midst of one of the busiest summer hurricane seasons in the Atlantic—one that marked a number of notable firsts, including 28 named storms, 15 hurricanes, 4 Category 5 storms, and 4 major hurricanes hitting the United States

(NOAA 2005)—Hurricane Katrina became one of the five deadliest storms to ever impact the United States and is also the costliest natural disaster in US history, having caused over $108 billion in damages (Knabb, Rhome, and Brown 2005). While Katrina demolished a huge stretch of the Gulf Coast, causing massive flooding and destruction from Texas to Mississippi, the technology usage in the hurricane's wake focused on finding missing persons. People flocked to various sites only to find the failures on CNN and the confusion on government and organizational websites. They leveraged the openness of the listings on Craigslist. This disaster was one where data were obscurely located and isolated; everyday people were unable to penetrate it. Without spaces to confirm information, people were never able to repurpose content as valid, useful knowledge in their communities.

London Bombings

On the morning of 7 July 2005 during rush hour, London's transportation system was the site of four coordinated suicide bombings. Known as the 7/7 bombings, 52 victims were killed and several hundred were injured by these terrorist attacks, which targeted the London Underground subway system and a double-decker bus (Intelligence and Security Committee 2006). Many of those affected by the attacks were carrying technologies that enabled them to take photos of the scenes of disaster and of the exodus from the city. This disaster saw a large spike in technology use, including the first video from a camera phone of a survivor escaping a terrorist bombing. People communicated via cell phone data streams and connected with others online. Social websites such as Flickr and mobile blogs became locations of information coordination as everyday people began to piece together information about this catastrophe.

Mumbai Attacks

Lasting several days, the Mumbai attacks of 2008 claimed the lives of 173 people and injured more than 300 (Press Information Bureau, Government of India 2008). Also known as the Mumbai bombings, these coordinated terrorist attacks began 26 November and continued until 29 November. Affecting multiple locations, including the subway system, the Taj Mahal Palace hotel, a women and children's hospital, a café frequented by tourists, and the Jewish community center Nariman House, these attacks rocked social web participants throughout the globe as they raced to help coordinate information about the attacks and locate the missing. In particular, the events at the Nariman House and the Taj Mahal Palace hotel captured public attention, and the media were often dependent on social web participants to relay information to those outside the affected area. Twitter became a site of great activity as participants used it as a form of citizens-band radio, notifying each other of affected areas and coordinating data collection about victims. The achievements of the participants and moderators across all of these systems were a major tipping point in the use of the social web in times of crisis.

Overview of Chapters

I structured this book to guide readers through the three phases of content transformation: data, information, and knowledge. I have organized the cases mainly along the timeline on which they occurred. The only exception to this order is the placement of Hurricane Katrina's case study before the London bombings case study. I have two explanations for this exception. The simple explanation is that in the Hurricane Katrina case study, content did not move beyond the data stage. I untangle the more complicated explanation further in these chapters, where I discuss issues of content control, gatekeeping, and digital literacy. While these technologies continue to evolve, we must step back to see where we started and to examine the reasons the social web has evolved to where it is today. By focusing on how content moves from data to knowledge, we can assess whether today's social web tools are living up to participants' expectations and needs.

This introductory chapter defines the concept of social web systems, discussing the need for an interdisciplinary worldview for architecting these technologies, and gives a brief overview of the major disasters surveyed in this book. Such an examination is, in itself, an interdisciplinary task, for any analysis of the building tools for the social web must take into account the many systems that mediate communication among people, organizations, and technologies. Discussing the need for shared perspectives across humanities and sciences, industry and academia, this book calls for research and practice that can aid in eliminating boundary areas. We need to move beyond traditional concepts of communication, networks, and participation.

Chapter 2 takes a deep dive into relevant literature focused on the theory and methods that I present. I discuss how experience architects can use actor-network theory (ANT) as a method for architecting effective systems of communication. Using ANT to analyze moments of exchange and to map the actors participating in networks is critical to understanding the context of these situations. Researching how people find and exchange information, share links, offer assistance, and so on is important for understanding how individuals can and do work to locate, validate, and distribute content. This framework, in turn, provides a mechanism for tracing these scenarios and tasks in order for architects to make visible the work of both the participants and users of mediated systems.

Chapters 3, 4, and 5 look at three distinct cases describing how everyday people use social web tools to communicate during times of disaster. These chapters illustrate how these disasters unfolded and examine these systems as participants tried to push content throughout the network.

Chapter 3 examines how everyday people and mainstream websites situate data. As the first stage of content, data are everywhere. Unfortunately, the case that I examine in this chapter is left in a perpetual state of data, unable to move forward because of gatekeeping, information mismatches, and misunderstandings following Hurricane Katrina. The key to creating participatory spaces where data can flourish is understanding what kinds of data are contextually useful. Discovering how participants locate data can lead architects to an understanding of how to build systems for improved findability and usability. In doing so, we can create systems that allow for improved flows of information, a difficult task as illustrated by these case studies. This chapter argues for more open systems that allow data to intermingle and coexist.

In chapter 4, I employ numerous case studies to illustrate how everyday participants in the social web work to verify and validate content across their networks. These cases illustrate how participants turned data into information, repurposing and linking people, websites, social networks, and various other sources and technologies during the London bombings. Observing and analyzing how individuals and technologies verify information is central to understanding how communicators can better research these interactions and design for these experiences. By looking specifically at high-pressure cases, researchers can see how participants use the tools that are available, and experience architects can envision new solutions for products and services.

In chapter 5, I analyze how participants redistributed content to the community as knowledge during the Mumbai attacks. Although this phase is the final stage for content translation, it is a major moment for moving information outward to people who need it most. After actors acquire data and various participants validate that data, that information is redistributed to the community as knowledge. By observing and analyzing these interactions, researchers can understand how social web systems work and experience architects can learn how to design for them.

Chapter 6 discusses the major issues surrounding the role of experience architects for researchers, practitioners, and educators. It also reviews recent disasters and the ways that these events have affected social web tools and their participants. In technical and professional communication literature, the site of study is typically the workplace. The social uses of technology, the communication issues of everyday experience, frequently go unnoticed by our discipline. However, these sites of everyday use and participation are rich with examples of how people communicate outside the authorities of large organizations, corporations, and many other institutions.

Certainly, the use of the social web during times of disaster is a largely untapped site of study—one this book addresses. The social web in general has received little attention compared with other, more traditional user-centered design topics. The use of these technologies has engaged a participatory culture, sometimes adding to and often circumventing current system implementations and traditional media channels. "Participatory culture" is the term used to describe how people are actively engaging with digital content, building networks, working across spaces, and connecting in productive ways. Participatory culture is not consumer culture in that participatory culture is about production and creation by everyday people rather than simply consumption. Participants share content, create mash-ups, and remix material from multiple sources.

In these definitions, I am partially paraphrasing Jenkins's rich work on the topic as well as looking at this movement from the perspective of experience architects. In Jenkins's (2012) white paper for the MacArthur Foundation, he states that "a participatory culture is also one in which members believe their contributions matter, and feel some degree of social connection with one another" (3). This definition works well for the phenomenon this book describes. People reach out, participating in new ways that surprise even those who build the digital spaces. They list missing persons in the lost-and-found section of Craigslist. They coordinate information validation across multiple comments at the end of news articles. Today, these kinds of participatory activities happen during disasters just as easily as they occur daily across the social web.

As the industry moves toward producing more participatory tools, we must reassess how to architect for experiences that must support information sharing, preferably in ways that are socially useful and contextualized. Participants want to change the lyrics, alter the colors, and build their own views. Architects need to determine ways to create experiences and determine the extent to which their systems should allow for participation. Within fluid ecosystems, participants are using multiple tools, data sets, and media, both off-line and online. We are past the point where we can consider only one system for a single user. Architecting for participation means that we must take into consideration the entire spectrum of interactions and prepare for fluidity of these experiences.

Who This Book Is For

This book is a call for a new kind of profession: experience architecture. Looking across the entire spectrum of humanists, technologists, and social scientists, this book seeks to encourage them to continue to branch out of their traditional roles and become more engaged in the research and design of these systems. Throughout this book, I intentionally refer to this work as experience architecture done by experience architects. But make no mistake—this work is ours as humanists and technologists: to extend our role as information designers and move into a more central role as experience architects, leading production and research teams in this role. Through these case studies, I outline ways in which we can listen and give more attention to the social web by venturing into these people-powered spaces where participants make do with the tools available. By spending time within these structures, tracing the actors, and participating in acts of communication, we can explore how these tools may evolve to better support new and innovative experiences.

In 2001, Hart-Davidson asked, "Why not us?" in an article discussing the technical communication field's move toward this work (146). We are well equipped to respond to this challenge of mediated design because of our deep tradition of user advocacy and empowerment design (Davis 2001; Grabill and Simmons 1998; Johnson 1998; Miller 1979; Mirel 1996). Our training in user-centered design and communication prepares us to sit at the center of project teams focused on tasks as diverse as usability analysis, project management, and clarity in design. The technical communicator is an "agent of social change" (Savage 2004, 183). This moment is when we shift our focus toward participating in and researching the design of interfaces, interactions, and processes and toward analyzing the use of these genres and associated tools (Potts and Jones 2011; Slattery 2007; Spinuzzi 2003, 2007; Swarts 2008). And all of these developments contribute to our ability to work effectively in complex contexts where we must continually assess and employ different social web tools to address changing situations.

This move is a major step toward moving the humanities, and especially the field of technical communication, away from being the "handmaiden to technology and science" and instead to "focus on rhetorical problems with a particular emphasis on domains of technical and scientific complexity" (Grabill 2009). Such an approach is imperative because today's social web experiences involve multiple systems, participants, and organizations. These are not only problems of interface and structure but problems of culture and participation. To put it more succinctly, we need to address the capacity statement for the field presented by Grabill during his 2009 presentation at the annual meeting of the Association of Teachers of Technical Writing: "The ability to identify, understand, and solve communication problems within organizations and in public space, with expertise in domains of technical and scientific complexity and with the ability to provide concrete solutions, including original research, strategy, user interfaces, and software" (Grabill 2009).

By becoming experience architects, we can help create systems that tap all possible means of collecting and exchanging up-to-the-minute, accurate information, and we can aid in the communication of knowledge among participants. In a disaster, this information contributes to the creation of a participant's ongoing narrative of the critical events that have taken place and of what is happening to the people involved. In everyday use, these narratives are more specific to participants' immediate needs of posting family photos, sharing information about their illnesses, or building costumes in an online game. Because we can solve only the problems we recognize and acknowledge, we need to possess tools with which we can locate the many technologies, people, groups, and websites through which people are sharing information.

The practical application of the methodologies that I discuss situates experience architects alongside everyday people as active contributors empowered to interact and participate with and within systems. This view moves away from building technologies for passive users who are stuck using whatever system is available. Architects must see exchanges firsthand in order to understand participants' experiences and learn how to architect within these systems (Potts 2009d).

For years, participatory designers have asked users to take part in design sessions. Now we must become active in these networks as a way to understand how people manage content with each other through technologies. Becoming participants allows us to experience firsthand the tactics and tools, the communities, and the elation and frustration felt by those victims, their family and friends, and complete strangers who turn to their social networks during times of crisis. And we can expand this method beyond disaster research into other areas in which experience architects can effect change.

We must now encourage the researchers and architects of our systems to, as Halloran (1982) put it, start "pulling oneself up by one's own bootstraps" and engage in laying the foundation for developing those "proper habits, and hence character" by carrying out "virtuous actions" (61). In this case, our task is to understand the landscape of possible opportunities—be they systems, groups, or people—before diving into designing individual tasks and interactions. Only by working with our participants can we create these systems. Because we cannot consistently predict the kinds of information that will be important to specific people in specific situations, we need methods to understand the dynamics between participants and commonly used technologies. These next chapters illustrate how to perform this bootstrapping and how to encourage our researchers, architects, and students to do the same.

We have a lot of work ahead of us if we are to build systems that are contextualized, robust, and appropriate. As researchers, we must engage with participants within these social spaces. We must make do with the tools available, learning how to improve them through observation of participants and our own use. As practitioners, we must look outside our own walled gardens and build for ecosystems. We must architect for flexibility, providing space for our participants to engage, to participate. Lawley (2004), in her post on researching the social web, expressed her hope that "as more researchers and scholars become users of these tools, rather than observers of them, that the range and depth of the research in this space will increase—as will its value outside the walls of the academy." The technological capacity and cultural interest to improve these experiences exist. Now is the time to discuss how we will move forward to find solutions as researchers and as experience architects.

2

METHODS FOR RESEARCHING AND ARCHITECTING THE SOCIAL WEB

In the popular use of social web tools such as Twitter, Instagram, Facebook, Pinterest, Flickr, and *Wikipedia,* we have witnessed a sea change between "use" and "participation." While the former is associated with years of user-centered design theories and methods, the latter is bringing forth new ideas, theories, and techniques on how best to research, build, and support sociotechnical systems. Unlike prior experiences in which users marched through a set of interfaces and stayed contained within systems, social web participants consider an entire ecosystem of solutions for communicating with others across multiple networks. As researchers and practitioners, we must reconsider our preconceived notions of "social" and explore these spaces as participants. Through these activities, we can begin to understand how to build more flexible systems, methods, and policies.

In this chapter, I explore these concepts further and begin a dialogue on new methods for building flexible systems that better support the experiences of social web participants. I map the shared connections among discourses and highlight their unique contributions partially with the aim of providing a pathway to a participatory view of experience. In taking this viewpoint, we can consider participation across an ecosystem rather than within one single website or application. This full consideration is experience architecture. A term meant to complement terms like "software architecture" and "user-experience design," "experience architecture" is becoming more common in the industry; it describes the role of architecting experiences across applications, websites, and services.

Answering a series of specific, communication-related questions is critical to understanding these exchanges so we can then understand the research issues, research questions, design issues, and design decisions that we must address before we recommend and make changes to designs and policies. By examining these moments of connection, we better understand how users locate and validate information and exchange knowledge. We can locate activity and discover who is participating in our networks. Below is a brief list of questions that have guided my own research and practice in disaster and everyday systems architecture:

- Who is locating data?
- Where do they locate the data?
- Why are they trying to locate the data?
- Who do they tell about the data?
- How do they tell others about the data?
- Where do they tell others about the data?
- How do they validate the data?
- Where do they validate the information?
- Who helps them validate the information?
- Who is charged with redistributing this knowledge?
- How do they redistribute this knowledge?

In case studies in later chapters, I use the theory and methods outlined in this chapter to explore questions like those listed here. I further discuss the application of these methods in those chapters. The examples are extreme, examining how people, technologies, and organizations leveraged social web tools during times of disaster. When disaster strikes, people reach out. Wanting to know the details of what happened, what caused it, and who was affected, people turn on their televisions, participate in online communities, and exchange information in person. Questions drive this process. People focus on finding basic information to provide context: What is going on? Who was affected? Was it anyone I know? If so, how do I contact them? When did this happen? And so on.

In outlining these methods for tracing digital culture, I draw on theory and methods from sociotechnical-systems theory, user-centered design, and technical communication to form frameworks. This discussion stems from the view that we cannot architect systems without participating within them. We need to move away from the typical researcher's stance of impartial observation. By participating, we will at the very least know the bugs in our systems. In the best-case scenarios, we will learn about the difficulties facing our participants. And these moves go beyond simply hanging out in our own systems; we need to work across ecosystems to see how and whether our own products, services, and policies are playing well with others. I discuss more of this debate in the next section. This discussion serves scholars and practitioners interested in learning more about how to research experiences. In turn, their research should lead to improved systems, policies, and architectures. To make these moves, we need to discuss the frameworks for this kind of analysis. By examining how content moves through multiple stages of validity and how networks form and become mobile, we can understand the context in which these moves are being made. We can also learn how to use frameworks in order to analyze spaces and research systems toward the goal of determining how information flows through the network. An analysis of this kind moves us closer to being able to build for experience rather than simply use.

Users and Participants

Members of many fields generally understand user-centered design as referring to creating products from the user's perspective as opposed to the software's perspective (Beyer and Holtzblatt 1998; Cooper, Reimann, and Cronin 2007; Garrett 2003; Krug 2006). Practitioners typically focus on different aspects of user-centered design, including conducting research, designing visuals, and developing software. The practitioners hold various titles, including "experience architects," "user-interface designers," "interaction designers," "information architects," "usability engineers," and "developers." Within the user-centered design methodology is a set of practices that encourages the producers of technology to understand the perspective of the persons who use the technology (i.e., the users) in order to develop task-based architectures that encourage ease of use (Courage and Baxter 2005; Hackos and Redish 1998).

While participatory design methods encourage users to contribute to design work (Bødker et al. 1987; Ehn 1989), designers use them to gain brief insights into how a given audience is using a particular item. Some critics of these methods say that participatory design is too limited in scope—that it looks at users rather than experiences and enables designers to disconnect from the users of the products they design (Beyer and Holtzblatt 1998; Forsythe 1999; Spinuzzi 2002). I propose that architects and researchers become the users of and participants in these systems. Only from such an up-close and engaged perspective—formed when individuals are active participants in the moment—can researchers understand what the connections between user and technology are and how such connections are made. This advancement is one step toward participant-centered design methods.

Another key factor in these discussions is one of terminology: "users" versus "participants." In looking specifically at the social web, Vander Wal (2007) emphasized the shift away from users and toward people, thus emphasizing participation over use. I take this idea further and focus primarily on participation and participants as they relate to people's activities in the social web. Early discussions of user-centered design emphasized one user working in one system. This work focused on activities taking place within a vacuum of computer-based interactions. Such an imagined vacuum lacks the surrounding social and off-line contexts that inform the user's activities. We are past the moment when we can ignore these ecosystems. Participants now work across multiple systems, balancing activities on multiple technologies, connecting to various applications and websites, and accessing spaces through a plethora of devices.

While some designers are able to "codesign the system with the users" (Beyer and Holtzblatt 1998, 370), product teams rarely have the opportunity to meet with, interact with, and learn from their customers. We are just now starting to emerge from a time when users were seen as victims and designers as saviors (Spinuzzi 2003). These heroic designers would save the day with new technologies to rescue the users. By building what they thought users needed, designers could magic up solutions without much investigation. But without this kind of user input, we have created systems and policies that are not serving our users. A more participant-centered investigation would involve researching, analyzing, and participating in the various technologies, people, environments, systems, and so forth that the users are engaging with to reach their goals. Internet-based technologies provide many opportunities to interact with participants online. We can and should collaborate with our participants because "the purpose of public discourse will not be to persuade but to participate in an ongoing exchange of ideas with other people and other cultures" (Zappen 2004, 161). Reaching beyond the tracing of a single-user experience, such research seeks to look across an entire ecosystem of technologies and people to understand the context of use and to support such use.

Researchers and practitioners begin to understand content, context, and users by participating in these spaces (see figure 2.1). Deploying methods lifted and modified from more traditional, anthropological approaches such as participant observation (DeWalt and DeWalt 1998; Ervin 2000; Stocking 1983), we can build a framework for participating in, analyzing, and mapping these spaces. Practitioners still do not consider contextually situated approaches to understanding the architecture process—such as researcher-practitioners embedding in the practices of the users or participating within networks and communities as active participants—to be typical practice. Such instances of embedding could include viewing YouTube videos and comment streams, observing the use of certain hashtags during earthquakes, validating links attached to various blog comments, and participating to provide information and make connections across these systems. Although contextual inquiry, a field-based method, and site visits are gaining traction in the industry (Courage and Baxter 2005; Potts and Bartocci 2009; Saffer 2009), we must spend more time participating before developing new structures or modifying existing ones.

While the disaster examples I present in this book are extreme, everyday practices are also filled with information-gathering tasks. People need answers, so they dig across systems to find them. To locate these answers, people need systems that can help them coordinate information. These systems can be made up of other people and can also include technologies. The systems we create, regardless of their intention, will be used during times of need. If we understand how to improve systems for these bleak disaster situations, we can certainly improve them for everyday use. While one size never fits all, the frameworks that are participant centered at least help us consider how we might be able to research these

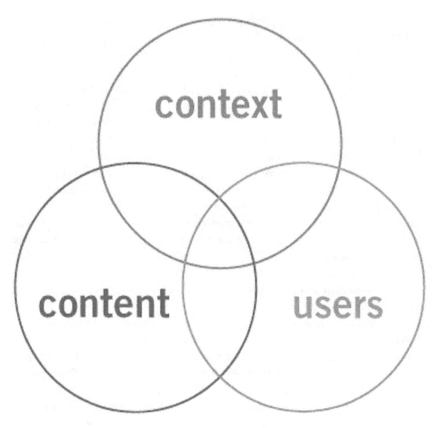

FIGURE 2.1 Understanding the intersections among content, context, and users (Morville 2004). Reprinted with permission from author.

spaces in ways that transform the results of our research into practical applications. We no longer have passive users (if we ever did). We must start building spaces that can engage participants; we must now evolve our methods for exploring experiences and building for flexibility in social web systems and policies.

Content and Exchange

In the social web, people participate in various ways, making connections and sharing content. Spaces are filled with different kinds of shared content, including status updates, links, images, videos, tags, memes, hashtags, tweets, and bookmarks. About 95% of Americans 18 to 29 years old and 81% of all American adults use the internet (Pew Internet and American Life Project 2012b). And they are doing so on laptops, desktops, and mobile devices such as smart phones and tablets. These forms of communication are changing rapidly in language style and in interface, and the context of use is shifting just as quickly. No longer trapped at their desktops, people are participating in ways we did not anticipate even five years ago. The future will bring many more ways to participate through these kinds of systems. As Burgess and Green (2009) recently asserted, "It is possible to imagine population-wide literacy in which everyone has the ability to contribute as well as consume" (131).

Nearly every week, a new online communication tool, service, or mobile app appears for us to explore. Here are some brief examples of this kind of use and this kind of content sharing:

- On a social network such as Facebook, a woman checks in to restaurants and tags her companions.
- On an information network such as Twitter, a fan retweets a link to a news article on women's soccer.
- On an image-sharing site such as Flickr, a student uploads an image from a study-abroad trip.
- On a free encyclopedia such as *Wikipedia,* a musician updates a page about notation.
- On an online pinboard such as Pinterest, a mother adds an image for a nursery.

This list is by no means comprehensive, but it illustrates example tasks, such as updating, tagging, linking, and communicating, in these various systems. People are accomplishing these tasks using a variety of devices, with mobile usage increasing. In the United States alone, recent figures show that 85% of adults are using cell phones (Pew Internet and American Life Project 2012a). Yes, we have certainly made gains in the physics of flying-bird apps. But we have also created systems that trap content in specific applications rather than allowing content to flow across multiple ecosystems. We are in another great moment for new technology, but we are unclear how much research is going into these boom-time applications, as was a major issue in the dot-com boom of the late 1990s. We now have to-do list applications that do not integrate with calendars, PDF annotation systems that do not allow comments to be viewed in other PDF-viewing applications, and other new walled gardens: systems that trap content within their own system, not allowing participants to share and exchange content with others. We have a continuous struggle to build systems that are not only useful but also culturally relevant to participants. To improve systems, processes, and policies, we can look at how people try to exchange content and spread information.

One method for tracing digital culture is to examine how participants exchange and transform content. In this section, I discuss these articulations in terms of defining content types, understanding contextual use, and tracing these content transformations. We can analyze the work of social web participants— both human and technological—to understand how they locate data, validate that data as information, and republish the data to the community as knowledge (see figure 2.2).

Data ⟶ Information ⟶ Knowledge

FIGURE 2.2 The process of locating data, validating that data as information, and redistributing the data as knowledge.

While information architects such as Morville (2005), internet studies scholars such as Weinberger (2012), and systems analysts such as Kock (2007) discuss these terms in different ways, I take a more rhetorical—and even a pragmatic—approach. I focus on participation (audience), events (exigency), and architecture (form and context). We must not only pinpoint how content moves through the three stages, but we must also understand the context in which this activity is taking place. Researchers and practitioners typically have strong opinions about what constitutes content and context; I briefly outline these concepts by defining the terms "data," "information," and "knowledge" as I use them in this book. These three key forms of content provide a method for examining and discussing systems and activities:

1. The initial form of content is data. Data can appear in networks as words, phrases, images, symbols, and so on. A simple example of data is a Twitter stream. Without any context, a Twitter stream is just data: links, text, usernames, and hashtags.
2. Information is the second phase of content. Information is validated data, and validation can come in several forms. For instance, participants can connect two pieces of data, such as an image and a name, to pinpoint a person's identity. Participants validating data creates richer, useful, and contextualized content.
3. Knowledge is the final stage of content. Knowledge is information that is shared within the network. It takes a form that allows for repurposing and distribution.

In the case study chapters, I discuss these three phases further. The phases are not rigid groupings so much as porous states through which content can move as details change. Throughout this book, I build upon these three stages, illustrating how participants transform content using social web tools. To understand the phases further, we can move on to discussing networks of activities and the actors that participate within spaces.

In examining the stages of content, we can see how everyday people create shared content. We must pinpoint these exchanges. Vander Wal (2007) states that "people express the richness with another person by a story about the person" (slide 18), highlighting a need to understand the design of social web as the mechanism for connecting people through these services. Moving through digital spaces, participants can construct and share their content across a social ecosystem made up of multiple sites, tools, and media. This content includes a plethora of types, including images, comments, status updates, links, chat sessions, and others. For this content to be useful to the participants, it must be able to move through the three phases in ways that can support these activities.

In other words, for systems to meet the needs of participants, they must be participatory. In defining participatory culture, Jenkins (2009, 5–6; 2012, 7) listed several specific attributes. Abbreviated, these key factors are as follows: low barriers to civic engagement, participant and producer support for content creation and sharing, informal expert mentorship for novices, belief among members that contributions matter, and social connection among members. The case studies in this book manifest these attributes. In some examples, the barriers for engagement are too high; in others, the social connections are what help propel the network forward. In many of these cases, little support for creating and sharing exist, but participants' belief that their contributions matter is what helps form the network in the first place. We have the technological capabilities to build for this kind of participation; we need to improve our architectures for participatory cultures.

To improve our architectures, we must look at moments of exchange among participants. A participant can be any actor within a given network—a person, technology, organization, website, document, event, and so on—or any combination of these actors. We find one example of this participant exchange activity on Twitter. Examining how participants retweet, why they retweet specific types of content, and how their networks use tags to categorize this information is important for architects trying to build systems, content managers trying to curate this material, and researchers trying to locate and follow these exchanges. In the next section on actor-network theory, I further describe actors and go into greater detail on participant exchange activity. For now, we must recognize that exchanges can occur between people and technologies.

Networks and Agency

Actor-network theory (ANT) originated in the work of Latour ([1987] 2003), Law (1992), and Callon (1986), and Mol and Law (1994) expanded on it. Working in the field of science and technology studies, Latour shook up then-current theories in sociology by suggesting all participants, whether they are human or nonhuman, have equal agency to affect any given situation. Referred to as "actors," these participants can be people or technologies. Furthermore, these actors can come together to form temporary networks led by a spokesperson, creating assemblages of relations specific to an individual act or broader event and forming a collective, referred to as an "actant" (Latour [1987] 2003). An actant is a network comprising any actors—cell phones, blogs, people, and so on—that have the ability to act and do act within the network. Examining actors and networks provides a broader understanding of the people, organizations, technologies, groups, places, and events (in other words, the nouns) important to these mediated systems.

Applying ANT to social web experiences is useful because it allows the architect to identify the network of participants, from the hardware technologies used to the websites accessed and the software deployed. Mapping the actors available to these participants allows us to create systems that allow for information to flow across these systems to support their use. For example, for a software team building business software, understanding the people, technologies, and organizations in a network allows the team to have a greater understanding of who is available, what tools are at their disposal, and what might affect their ability to get work done. Knowing that certain tools are used in certain situations, knowing which people are sources of knowledge, and visualizing how all of them come together to get work done are vital sources of information for architects. With ANT, researchers can look across the networks of technologies and people to identify and understand communication exchange. This approach provides a new method for architecting social web tools and systems in order to support the work of participants. Tracing how actors form networks to exchange content is an important part of this framework. Being able to pinpoint the many stages of that process is essential to understanding how this work takes place.

At their most basic level, social web tools connect people through common tasks, scenarios, and activities. In architecting these systems, practitioners should "think of social software as software that connects people through activities" (Torkington 2006). Such a view aids architects in thinking about these connections in ways that move beyond the walled gardens of the past. Of course, this view is predicated on the ownership of information. I discuss such issues of agency and control in the case studies. To be clear, social web participants want content mobility; they want to be able to move and make use of content across the network.

Actors, Networks

For the purposes of this book, I assign the term "actor" to any active participants in the network. Actors can include a multitude of nouns, such as people, organizations, events, and technologies (e.g., devices, websites, software). These actors come together as part of a network to accomplish certain tasks. Interconnecting participants, a network is a mechanism for information coordination and the fluid movement of information. The internet is a distributed network[1] in that "there is no hierarchy between nodes, they are strongly symmetrical, and there are no predetermined routes between nodes" (Cantoni and Tardini 2006, 27). That flexibility allows for a space in which actors can come together and later disband.

Networks like those crisscrossing the social web are mechanisms for the coordination and dissemination of information. Identifying networks help locate the actors (nouns) in a network, which is required for researchers to help make sense of the many activities (verbs) that we can pinpoint. The use of ANT recently received attention from other technical communication researchers (Johnson-Eilola 2005; Spinuzzi 2003, 2008; Swarts 2010). Using this theory as a method gives experience architects and researchers a tool to identify actors in order to trace how these actors create information from raw data around them to meet their localized literacy needs. This move is a turn away from observing the use of static objects and is instead a call to build flexible tools that can accommodate the means by which participants exchange content. Such tools are active, organic communication tools that rely upon the input of these participants.

Spaces that support change also better support new exchanges, allowing actors to form connections that can wrap context around information. Law and Mol (2003) call such environments "fire spaces," spaces where connections among actors remain relatively stable while they add information to the network or modify it as content becomes highly mobile and sometimes unpredictable. For example, Flickr is a fire space for amateur photographers. Last.fm is for music fans, and deviantART is for visual designers and artists. For researchers and architects, understanding the social web as a fire space in which people use mutable mobiles (discussed below) to distribute information across the network is key. This concept forces researchers to approach different actor networks on the networks' own terms in order to understand how participants define and practice stability, particularly within the shifting contexts of the social web. Researchers and architects must become active participants within the communities and situations for which they architect social web systems. To map networks and trace the relationships among actors, they can apply a method based on ANT that accounts for the broader ecosystem of participating actors and traces the actors' relationships to one another. This method involves three stages and focuses primarily on the nouns in these networks and the relationships among them. In the following sections, I describe how architects can use this method while highlighting the mobility of information and the nature of relationships among actors.

Movement in Fire Spaces

The major tenet of ANT is that all actors have equal agency, regardless of whether these actors are people or technologies. While I am not suggesting a computer has agency equal to that of a person, examining actors together is useful for understanding the context in which these digital experiences take place.[2] Actors come together to form a network, creating relations that are often temporary and specific to an individual action or a broader event. Understanding this concept allows architects to see a landscape of participants—human or technological—who cooperate to create, validate, and share content.

As participants communicate across networks, they push information forward from actor to actor. The vessel that contains this content is referred to as either an immutable mobile or a mutable mobile, depending on the type of actor participation (Law and Mol 2003). In situations in which participants can modify the mobile—the vessel containing the communication—the mobile is mutable. When participants cannot modify the mobile, the mobile is immutable. Based on this concept, most of the activity on the internet centers on mutable mobiles; inference is clearly the case for texts created using social web tools, where participation is a key interaction point. Flourishing mutable mobiles enable new participant formations, interactions, and activity.

Architects must recognize these new formations in order to locate spaces for participatory practice that they can incorporate into new versions of social web tools. Social web tools also match the description of a fire space, where relationships are stable and the transportation of information is fluid, though unpredictable, allowing for others to modify and add to information as it flows across the network (Law and Mol 2003). Stability emerges from this fluidity—at first glance a contradiction. But the flexibility of social web tools allows actors to stabilize their networks. This concept of stability is essential for exchanging information in fire spaces, even when "constancy is produced in abrupt and discontinuous movements" (Law and Mol 2003, 7). Stability requires actors to conduct activities that reinforce the network, such as repeatedly sharing information (e.g., hashtags on Twitter) and connecting actors to each other by linking content. Both activities are tangible examples of stability within social web tools.

Formation and Mobilization

Participants in the social web are what actor-network scholars would refer to as "translators" who perform "translations." In the case of the social web, participants translate data into information—and then knowledge—through an extended process of network formation and mobilization. They punctualize (form) the network by moving through the stages listed here. We can map out this process by examining the exchanges between actors. Callon (1986) defined these four stages of translation in great detail:

1. Problematization is the first moment of translation. During this stage, volunteer actors rush to establish and define the event. They become anchor actors, central to the network, and volunteer to serve as obligatory passage points for content collection, validation, and distribution.
2. Interessement is the second moment of translation. During this stage, the anchor actors encourage participants to accept the network definition that they originally defined. At this point, the network becomes more centralized.
3. Enrollment is the third moment of translation. During this stage, actors align themselves with the anchor actors. In doing so, participants accept the

definition of the network. Their acceptance reinforces the anchor actor as being the obligatory passage point.

4. Mobilization is the fourth and final moment of translation. At this moment, all of the actors assemble across the network and mobilize to validate and distribute content. During this moment, anchor actors play an important role, aiding in the organization and distribution of materials.

Through this translation process, actor networks gain other participants who are also willing to translate more material into information, helping to form a group of active participants. When these participants tag a particular photo, they are bound by the prescriptions—the interfaces and interactions—of the system. The idea is to pinpoint actors in a given network rather than to overdocument a system and become buried in data not relevant to the experience. Participants then apply their own inscription to that photo by attributing a semantic value to it. They thereby turn the photo into an intermediary that can then help in constructing the narrative for the event (Latour [1987] 2003). This inscription repurposes the image for others within that culture, validating its importance to those affected by that particular event. By giving significance to the artifact, they are able to translate material for their community, leading to punctualization that creates a whole that is greater than the sum of its parts.

While moving across these networks, actors leave various inscriptions behind. Inscriptions can include tweets, posts, tags, and other types of content. Looking specifically at examples in this book, we can see that such traces of movements include comments posted to blogs and tags added to photos. Such traces help the developers of these technologies know how people are deploying and using their prescriptions for these systems in the field. When a system's prescriptions meet up against the inscriptions of participant use, designers have an opportunity to improve communication networks. To explore this idea further, we can begin to map these networks.

Identifying and Mapping

ANT enables us to robustly analyze and discuss the efforts of multiple actors to push information through a given network. Within this space, we can trace actors to give researchers and practitioners a more full picture of who uses these technologies, what tools are at their disposal, and what organizations may also have a stake in this use. Being able to trace information as it moves through these networks is key to mapping these actors and architecting for the smoother transference of information to support the use of these systems. While such mapping may not have been the original intention of Latour's work, this application is one way in which we can extend ANT to make it a useful method for researchers and practitioners.

From this point, the architect can pull back to observe and engage the entire network of actors (i.e., all the individuals interconnected by or interacting through the technology), mapping out the connections among systems, technologies, and people. This broader view bolsters the knowledge that architects gain as they participate in specific instances of exchange (i.e., moments when two or more actors come together temporarily to accomplish a task), embedded in the contexts that those who regularly participate in such ecosystems face. Mapping the participants and technologies involved in an organic, fluctuating system allows architects to understand the entire ecosystem that affects these interactions, off-line and online. In later case studies, I illustrate these methods in greater detail.

Within the cycle of software development, mapping is a key activity for generating everything from technical manuals to product websites to mobile applications. Software architects document states, processes, and technologies. For example, when creating a diagram for a new product design, software developers often create activity flowcharts. Because those designing social web tools must consider the many technologies, people, and organizations involved in these ecologies, the examples in this book illustrate a new method for diagramming these participants based on ANT. Such diagramming can aid in the development of complex social systems as well as the development of everyday business applications. For experience architects, such approaches have a number of distinct advantages. First and foremost, they allow experience architects to understand the scope of the product—who will use the product, what technologies will be at the participants' disposal, what organizations may be involved, what sort of events will be occurring, and so on.

The following sections explain how to map these networks by visualizing the basic context in which participants work. With this framework, we investigate the moments of exchange among system participants as they interact with one another. We explore, document, and visualize the networks of participation. Coupling these methods provides a framework that researchers and practitioners can use to examine the issues in these situations. Creating visualizations that describe actors increases our ability to be stronger user advocates and project leaders. With this knowledge, experience architects can further research these actors, provide information for product teams and marketers, and be a part of building a more contextualized experience for the network's participants.

Traditional Software Diagramming

Traditional diagrams in software engineering focus on a specific system's possible responses to a user's actions. Thus, these diagrams are often system-oriented visualizations that capture a single participant's use of a system. These flowcharts, timelines, and assorted diagrams tend to follow an actor employing a single system to complete an individual task. The original developers did not intend for these diagrams to illustrate the interactions of multiple participants across multiple technologies in a social web ecosystem. None of the diagrams impart an understanding of an ecosystem of actors. These drawings do not contain any contextual clues about what kinds of actors are supporting these scenarios, tasks, or system statuses. Rather, these diagrams only reveal what is present in a particular system for a particular user or technology.

For such applications to be successful, the entire product team must be aware of the major people, organizations, technologies, events, and so forth in a system before any development work begins. Such situations can be prime opportunities for experience architects to contribute to project teams by providing team members with essential information. In working through these maps over several projects, I have found ANT mapping to be useful in extending the view of networks to the many technologies, people, and organizations that extend the traditional boundaries of user experience (Potts 2009a, 2009b). While traditional software diagramming methods focus primarily on the process flows and system states for specific tasks and technologies (Fowler 2003), initial ANT attempts called for development teams to think globally about architecting experiences not bound to a single system. Considering the technologies, websites, users, and organizations involved in each scenario enables practitioners to build for these experiences (Potts 2008). This particular method is apt because participants use a multitude of technologies, conversing with numerous people and shifting their use patterns to fit their current tasks.

Besides using ANT to examine how information surfaces and temporary connections are made, we can also use it to trace the list of possible actors and associations within a network. Any actor that is an active participant—a noun—in the network is relevant for inclusion in such maps if and only if that actor's inclusion is mandatory for the central object of study to exist. Such central objects can include blog posts, hashtags, photos, and other content. In the workplace, people can include customers, websites can include a company's site, and events can include product launches. By diagramming actors, we also identify the context in which experiences exist. These kinds of diagrams can help the team understand where actions intersect and diverge, improving on more traditional software engineering diagrams.

Building ANT Maps

Practitioners and researchers new to these processes might wish to begin applying these concepts through inadequate approaches. A team should develop ANT maps after the team members have explored the network and pinpointed the actors in the system. Team members should begin by creating simple ANT diagrams that capture the people, technologies, organizations, events, and other relevant actors. Using the map as a talking point for the product team, they can clarify the context in which their participants will experience the system. For example, we might take simple objects, such as spreadsheets used across product teams, to discuss what actors must exist in that network in order for that spreadsheet to be an actor. As the diagram matures, teams can create stencils (discussed below) to show commonality among different actors. As the product team discusses issues of participatory experience, clarity can grow about the relationships among actors. Walking through these maps and conducting more research can reveal how relationships among actors are situated; these relationships could be tenuous, strong, ambivalent, essential, and so forth. Employing these methods to discover user needs and participatory contexts allows experience architects to lead these exercises and have greater influence on experience and policy. By participating in spaces typically thought of as the realm of software engineers, program managers, and policy makers, experience architects can be participant-focused advocates. Such advocacy leads to building systems that are more appropriate for helping to manage activities or even for advocating for new policies and processes.

ANT diagrams help scholars and product teams pinpoint the people, organizations, technologies, places, and events that affect the network structure. Visualizing networks in a simple, clean diagram helps in transferring understanding from one group to another, whether academic researchers or industry experience architects. Such a diagram enables research teams, product teams, and policy makers to better understand the people, organizations, locations, and technologies. By pointing to specific actors, we as researchers can visualize these ecosystems—the organic, engaging, and interactive networks. And in so doing, we as communicators can make more informed decisions about how we architect experiences for participation. Such work ranges from the most basic of interaction choices, such as where to place buttons on screens, to more complex systems decisions, such as usage policies. Building ANT diagrams involves three stages:

1. Pinpointing the central actor or object
2. Identifying relevant people, places, and things (nouns)
3. Weighing relationships

I discuss these stages here and showcase various examples of them via diagrams in later case studies.

Figure 2.3 illustrates a basic ANT diagram. To begin the diagramming process, we pinpoint the actor or artifact, such as a blog post, that is the center of activity. This actor stands as the central artifact of the activity being studied. For example, to determine the central actor when mapping participants tweeting about a hurricane, researchers would look at actors, including hashtags, participants, moderators, hurricane warnings, links to websites such as weather reporting sites, and third-party tools such as TweetDeck (Potts and Jones 2011).

After deciding on the central actor, the next step is to list all *active* participants that exist in the network. Obviously, the detail of these diagrams can become incredibly microscopic—including minutiae such as internet connections and

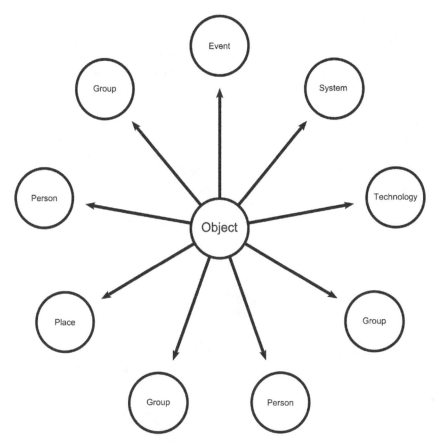

FIGURE 2.3 A basic actor-network diagram.

power cables—which is why we must focus on the concept of activity in defining these actors. The central actor exists because of support from the actors around it; if any single supporting actor is not present, the central object cannot exist.

In this diagram, the central actor is linked to various actors that are present because they directly affect the central actor; without these supporting actors, the central actor could not exist without any relevant people, technologies, organizations, and so on, to construct, make relevant, or validate this actor. For example, an image from a bombing cannot exist without the bombing itself also existing. The level of detail—such as listing a Twitter versus listing a specific Twitter hashtag—depends on the types of information that the researcher or architect needs to know about that network or actor.

These diagrams begin to form a sketch of the ecosystem in which the actor exists. By documenting these nouns, the team can see the current state of the network and begin to see who or what any new interfaces, processes, or systems will affect. Understanding these networks of participation is critical to architecting these experiences. Visualizing the people, organizations, and technologies involved in these experiences helps interdisciplinary teams share a common understanding about the context in which participants will approach these experiences. Such understandings are critical for the development of participant-centered experiences because they support the ecosystems that these participants are attempting to communicate across.

Next, we begin labeling these nouns by using unique stencils—example images that are meant to represent different actor types—recognizable to anyone who will be examining and using these diagrams (see figure 2.4). Again, we are focused here on identifying only nouns relevant to whatever network is supporting the central actor (Potts 2008). We use verbs later in the flowchart diagrams typically found in the software development process. We create these

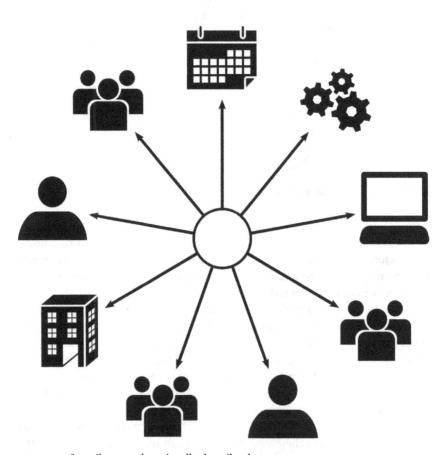

FIGURE 2.4 Stencils created to visually describe the actors.

diagrams after we have identified the nouns that will be affected by the technologies, processes, and services that will be invented for, modified in, or removed from the network. First, we must know the participants. Then, we can diagram activities and processes.

The stencils in figure 2.4 are not comprehensive. Rather, different situations require the use of different kinds of stencils. For example, a situation like a terrorist attack requires a stencil that looks like a bomb, while a situation like a new movie release might require a stencil that looks like a film. We should create stencils in ways that allow experience architects and researchers to understand and examine what structures are in place that can help permit or enable these experiences (Potts 2009a). Creating a common understanding of networks is key. Within this context, ANT maps are meant as a starting point for conversations. For experience architects, this process is important because of their role as participant advocates: they are tasked with locating, understanding, and responding to the needs of people who work within the interfaces, systems, and networks that any products and services are a part of. By mapping out the relevant people, organizations, technologies, and events, researchers and architects can better address questions of resources, use, and activity.

Diagrams can also visualize shifts in cultural practices, such as using Twitter to organize events, posting missing-persons pictures on Flickr, or validating content in a Google Doc. Such mapping extends ANT's vision of distributed agency and, by doing so, allows architects, legal experts, and policy makers to view the actors as participants within networks rather than simply as users. Within these visualizations, we can discuss the architecture of systems, processes, and policies to support human work in a given scenario. When we list relevant actors, we begin to sort out who is acting and consider their motivations in these systems (Potts 2009b).

Figure 2.4 shows many actors that support the network. By finding patterns across these actors, experience architects can create stencils to communicate associations. For example, the pattern of groups helps describe organizations and differentiate them from individuals of importance within the ecosystem. Note that this diagram lists human and nonhuman actors, pointing to the sociotechnical network in which these actors reside and allowing us to ask questions such as these: what ecologies can these actors leverage, and what groups have a stake in these actors?

Once we visualize nouns and see patterns emerge, we can extend diagrams by visualizing shifts in practice or even in types of relationships among actors. We can catalogue shifts from a number of different perspectives, including strength of ties, length of time, and history of use. When examining the use of social web tools, we can choose to measure the strength of relationships relative to time (Potts 2008). For example, if actors spend a lot of time exchanging information, then the lines connecting the actors would be thicker. If the relationships seem more critical due to the importance of the information flow, then those lines would also be thicker. The less time spent exchanging information and the less important the relationship, the thinner the lines.

A valuable use of these lines is to show relationships among actors across the social web (see figure 2.5). Tracing these associations makes visible the connections actors make across ecosystems. Within this context, examining issues in relation to the timeline or importance of relationships among actors shows experience architects, systems engineers, and researchers how information flows across actor networks. With this knowledge, we can architect systems, processes, and technologies that can mediate these interactions appropriately. Understanding these relationships helps us become better participant advocates, for it allows us to visualize the many people, organizations, technologies, locations, and events that are part of the networks for which we are building experiences. Mapping helps visualize the context within which these interactions

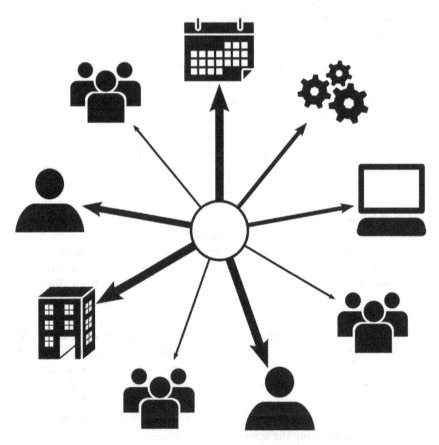

FIGURE 2.5 An actor-network diagram visualizing relationships through line connectors.

take place; thus, it provides common ground for discussions across projects and product teams.

The process of visualizing a noun involves identifying it within the network and creating a stencil for it. This approach is an essential one in relation to effective mapping using ANT. The idea is that once the nouns are visualized and patterns emerge, researchers and practitioners will have another way to extend these diagrams by visualizing shifts in practice. This option provides the benefit of revealing similarities in available participants, resources, and locations. Such advantages multiply when we use this approach to understand the social web.

ANT as a Method for Mapping Ecosystems

Taking up ANT can empower architects to map out the entire digital network of an ecosystem. In doing so, we can examine these exchanges on a much larger scale, seeing the networks as more comprehensive and complicated systems that require multiple actors to support these holistic experiences. ANT allows architects to map the connections among multiple sites to examine how people and systems are part of a larger network working to validate information. In tracing usage, we can discuss how agency fluctuates between the human and nonhuman entities, making the case for a more flexible system that can support these fire spaces.

By implementing ANT as a way to map these ecosystems, we can visualize the context in which these participants are communicating, thus discovering how sociotechnical systems work (Potts 2009c) and how they can affect policy (Potts 2010). As researchers, we must understand what exists in these networks in order to study and contextualize them. As communicators, we must architect for the context in which these experiences take place. Looking across networks and documenting the technologies, organizations, and people involved, we can create a better map of what is or will be affected by any further implementation of technology, process, or policy. Using ANT diagrams and their associated stencils allows communicators and scholars to make visible the people and technologies that participate in these networks. In making communities visible, we can more readily see the participant experience issues. These diagrams can help influence stakeholders in the development of more effective and efficient systems based on participation rather than use, allowing us to plan for appropriate solutions to the problems affecting the usability of social web experiences.

Using ANT can help us imagine ways in which the methods and theories we develop transcend our own professional community and are useful for those in other knowledge-making areas. By making such moves, we can continue our long-established tradition of being user advocates and become participant advocates. As Dubinsky (2004) states, "The best judges of the *making* are not the makers but the users" (5; italics in the original). We can take this concept further and say the participants and our participation will help guide us as we improve products, services, and policies for the social web.

Conclusion

This chapter covers discourses within science and technology, user-centered design, and technical communication. These approaches each contribute a key element to developing a new methodology based on cultural use, distributed agency, and participatory experience architecture. This method supports the architecting of experiences across ecosystems. Johnson (1997) characterized this new conception of audience as one that empowers the audience to be active and engaged: "In contrast to the addressed or invoked models of audience, the involved audience is an actual participant in the writing process who creates knowledge and determines much of the content of the discourse" (363). Clearly, we must engage participants in order to inform the architecture of these systems.

Spolsky (2004) argued that our prior understanding of and models for usability fail when it comes to assessing and developing social web tools. He explained that social software "mediates between *people,* not between the human and the computer." In explaining this distinction further, he stated that software developers must take into account sociology and anthropology when developing social web ecosystems: "After you get the usability right, you have to get the social interface right. And the social interface is *more important*" (Spolsky 2004; italics in the original). The methods described in this book emphasize the social—the building of systems to enable and empower human communication. Clearly, we need interdisciplinary work to explore methods that can help researchers and experience architectures to develop social web tools and policies that can react to the needs of participants.

Tools and humans are in continual flux; our social and political systems are equally in flux. We need to understand in equal measure that "the way we experience the message is shaped by the medium. And the ways we define information are shaped by the properties of that medium and the context of use" (Morville 2005, 48). Both systems and people are unbalanced due in this context to the often-unpredictable circumstances presented by modern life. What is predictable is that people want to communicate and that they often do this communication through the use of social web tools. Research and practical application has shown not only that this prediction is technologically and humanly possible but also that it is already present in its most rudimentary form online. In talking about this shift in culture, Jenkins (2006) stated that audiences "are demanding the right to participate within the culture" and that "producers who fail to make their peace with this new participatory culture will face declining goodwill and diminished revenues" (24). We are in the midst of this moment in experience architecture.

The next several chapters cite specific examples that show how systems have managed to restrict, empower, hobble, and enable actors, and how we architects and researchers can pinpoint network activity, map these networks, uncover how participants verify information, and determine the ways that actors transfer knowledge across systems. In the following case studies, I couple practitioner insights with academic research. In so doing, I bring an awareness of the ad-hoc, participatory activity of actors so that architects can improve how they develop and maintain these systems and so that researchers can delve into these issues and thus bring to light the issues that we have yet to discover.

3

LOCATING DATA IN THE AFTERMATH OF HURRICANE KATRINA

In August 2005, Hurricane Katrina slammed into the United States, devastating the Gulf Coast and displacing hundreds of thousands of residents from their homes and family members. Many people turned to the internet to locate missing family members and reconnect with those who relocated across the United States, and many of these people faced similar problems prevalent during other disasters. They attempted to use systems lacking participation, systems that did not understand the context of use, and systems that did not enable actors to locate information that would have made these exchanges more productive.

As Hurricane Katrina smashed into the Gulf Coast, the combination of a woefully slow and ill-equipped government response and walled gardens that the mainstream media built did not provide spaces for the participation of everyday people. Yet, this case study shows how people-powered, decentralized communication systems could eventually become effective during times of disaster. These media organizations, along with relief organizations and grassroots online communities, attempted to organize missing-persons lists, coordinate relief supplies, and exchange news about the situation. While Hurricane Katrina wiped out a huge stretch of the Gulf Coast by causing massive flooding and destruction from Texas to Alabama, people flocked to various internet sites only to find a cacophony of information and confusion. Activity erupted and died down on various sites, depending on how well participants could interact within these spaces. By using actor-network theory (ANT), experience architects can trace data as the data moves through these spaces, watching participants attempt to locate missing persons online. Though the social web as we know it today was still rather new and underdeveloped in 2005, these digital spaces often proved far more agile, adept, and successful at helping to reestablish communication for victims, their friends, and their relatives.

In mainstream spaces, hampered data flow caused difficulty with people's ability to participate in the network. Because the participants could not validate content, the data could not reach a stage where people could consider that data to be information. As such, the content was not recirculated as knowledge to the community. As the examples in this chapter show, the time in which Hurricane Katrina occurred was one when mainstream media were trying to own the conversation and keep participants on the periphery. At the same time, social websites were experimenting with giving participants more freedom to create and distribute content. The system failures on CNN's website were typical of a walled-garden system, one in which the data stay captive in a single digital space. Similar problems plagued the Federal Emergency Management Agency (FEMA) and the Red Cross websites, where these organizations locked down and controlled content. However, on Craigslist, the lost and found category of the New Orleans section was more participatory but often chaotic. While many of these attempts were well intended, none of these systems used a coherent model for participants to actively engage with the content. This lack of engagement made it incredibly difficult to transform data into valid information. Where and how could participants confirm or deny the posted data? While none of these system solutions were ideal, the examples in this chapter illustrate the push and pull for control of participation. In determining how participants worked around these gated systems, we can learn about how their engagement altered the very purpose of the systems.

In this chapter, I focus on the ways participants work to locate relevant content in their attempts to locate missing persons and share in data gathering efforts. Working to discover how participants locate relevant data is useful to understanding how we can build social web systems that support findability and help knowledge work flow throughout an ecosystem. In this chapter's examples, participants faced the same daunting problem as those who responded through the social web to the Indian Ocean earthquake:[1] encountering data disconnected from the contexts familiar to the participants. More so, in these examples, participants faced closed systems that did not encourage their participation. Arguing for more open systems that allow data to travel without such slippage, this chapter presents examples of the usability of such systems and discusses how to make improvements that would affect today's implementations. For our systems to be successful and useful, we need to reconnect content and context. Such reconnecting, however, is a serious challenge that will require collaborations across the sciences and humanities, as one can engineer the artificial and the other can contextualize the intelligence. Together, we can solve these problems.

Ecosystems and Data

For the purposes of this book, I define "data" as content found in the social web. This content can include status updates, tags, tweets, photos, videos, forum posts, blog comments, and other material that exists across these ecosystems.

Independently, data lack context. Without knowing more information, seeing a tweet that says "huge #blasts tonight" does not indicate whether this event is an act of terrorism or simply a party. Without any information about location, we cannot verify where this event is occurring. Without knowing who sent the tweet, we do not know whether this issue is serious or entertaining. Unless you trace this hashtag, you will not know if other participants are also talking about the event. Only through validation by participants and technologies can data become more useful and thus move into a state where we can label the data as information.

Participants in the social web often use tools to search for and locate data within these ecosystems. In particular, they use tools such as Really Simple Syndication (RSS) feeds, search engines, and other systems to collect and aggregate data from around the internet. Calling up specific content in their networks, users are able to use tools to ingest this data from various systems. Note my use of the term "users" instead of "participants" in this section. With this change I intend to point out just how nonparticipatory these activities can be. Although users actively search and seek out material on Google and other systems, these activities typically do not call for participation on their part. Users cannot catalog and curate; they can only receive search results and RSS feeds. They can only use, rather than participate, in these systems—which is the problem.

Google and other search engines are not effective for locating new content, especially live event content. This problem occurs because search algorithms function by rewarding established content rather than helping users find new content relevant to current issues. While you can search for and locate news reports, you cannot currently find hot spots of activity during these disasters. Finding these hot spots is simply out of context for the current purpose of these search engines. On the day of Hurricane Katrina, Google could not have known how relevant the content on Craigslist, LiveJournal, and specific blogs was to this disaster. While Google is making inroads to locating data, still massive problems persist in locating context-specific data that make an impact on these communities.

Google and other search engines are pull technologies, while RSS is a push technology. Data move from the source (such as a news website or blog) to participants at the participants' request. This information typically exists out of context in a reader of some sort, such as Google Reader. The system gathers all of this data for users and passes the data to them in the form of RSS, and the users can then read (or ignore) this content in their reader. Some web browsers such as Firefox allow people to subscribe to live RSS feeds that present new data as data emerge. Examples of these feeds include *BBC News*, sports scores, and blog updates. Users can then click on these live bookmarks and read the content on these websites. Many websites also allow the publishing of participatory content into an RSS feed. In all of these cases, users must follow specific protocols and preset interactions of either the feed reader or the browser. Of course, you would need to know ahead of time what RSS feeds you need to subscribe to in order for these feeds to be effective during a disaster. Such prescience rarely happens given that the sites of knowledge making are as diverse as the disasters themselves. While we do know that people are more likely to use technologies where their networks already exist (e.g., Facebook, mainstream media sites), where exactly this content will erupt is unclear.

In sum, in these systems, emergent data flowing in real time from disaster situations often slip through the nets that these tools cast. RSS feeds depend on already-established subscriptions to already-constructed websites. Search engines generally rely on keywords and well-established metadata structures. Such systems are not effective methods for tracing data as they emerge in a fire space. While much activity occurs over social websites that are familiar to everyday people, the groups, tags, and associated content are not easily identifiable to members outside the various communities that develop them. And even when participants can find these locations of activity, as cases in this book illustrate, locating these data is only the first step to coordinating information with others.

Data Mobility

An actor network comprises human and nonhuman actors. Human actors include everyday people, system administrators, government officials, and anyone else who acts as an individual in this network. Nonhuman actors include technologies. They also include organizations such as the Red Cross, the government, and events such as bombings, hurricanes, tsunamis, and earthquakes. Understanding how data move through these networks is critical to analyzing how participants work within these systems. Locating the spaces in which this work takes place can help us research this activity and improve these experiences.

Latour (1999b) uses the concept of the black box when discussing actor networks that contain other networks within them. One network appears as a unified system but in reality hides other networks operating just out of sight. These are networks within networks. Participants in the social web dig through these black boxes as best they can, trying to locate data that can help them solve whatever tasks they are trying to complete. For example, when trying to track a disaster on Twitter, knowing what hashtags people are using is important. When trying to track a disaster on Flickr, knowing what tags people are using to describe images from the event is imperative. Once actors open black boxes, they can then pull data out of them. The problem is in knowing where to locate these black boxes. For data to flow through these ecosystems, the technologies must be open and flexible enough to allow for movement across these systems. As the Katrina case shows, we were a long way from this kind of flexibility and flow in 2005.

A key to understanding social web architectures is the concept of immutable mobiles (Law and Mol 2003). Immutable mobiles are objects that have structure and are unchangeable. An example of an immutable mobile is a hashtag that has become stable across Twitter, such as those for conferences and television shows. The relationship between separate immutable mobiles never changes. These relationships remain constant and stable. Changing hashtags often leads to a fracturing of the discussion across Twitter because people may use multiple hashtags to discuss the same phenomenon. Mol and Law (1994) argue that all networks comprise immutable mobiles. They explain that if these mobiles ever become mutable (changeable), then the network will fall apart because meaning is lost and cannot be recovered within the same network structure. For example, if conference goers move from using #sigdoc to #sigdoc12, they may lose conversations between groups using one hashtag over the other. The lack of connection between the two hashtags can cause major communication problems during the conference. As a more serious example, during a disaster, people have difficulty communicating because of these hashtag confusions, as seen during the 2010 earthquake disaster in New Zealand where multiple hashtags were popular (Potts et al. 2011). Different groups on Twitter used various hashtags, such as #eqnz, #earthquake, #quakenz, #nzquake, #nz, and #christchurch. Across these hashtags, participants shared useful information, but they also duplicated efforts and simultaneously lost information (Seitzinger 2010).

Black boxes that appear whole can turn mutables into immutables, or at least foster a stable appearance. Locating these stable mobiles is key to uncovering activity in the social web and helping participants turn data into information. The case presented later in this chapter discusses the black box created on CNN's website, a site that creates strong boundaries between the organization and the audience. This boundary is a walled garden that protects the immutable mobile from outside participation or even interference; only official representatives within CNN's system can make changes to this content. Moving immutable mobiles throughout the system, actors work to share this content. These data move throughout the system as actors tag, copy, and link to this content. Immutable mobiles can retain their power for long periods of time primarily because of the cooperation among actors, and the same is true for mutable mobiles. In the case of an immutable mobile, the network is often stable to content producers. If this content could be mutable, then participants could help turn this data into information. The question, then, is this: how can we design for participation while still having stability in these systems? Architects must account for actions that disrupt the immutable mobiles, especially actions that aim to destabilize these networks. As this movement accelerates or takes hold in specific systems, the need for stability increases. In the case of disaster, such activity tends to accelerate quickly, and this situation becomes a critical problem. Eventually, this activity must stabilize somewhat for data to eventually move into the information stage, where we can validate and contextualize them.

Fire Spaces

Social web tools can erupt into what Law and Mol (2003) refer to as fire spaces. These are spaces where movement is kinetic, unpredictable, and often disruptive. Within a fire space, new data affects actors in ways to which they must react. These reactions result in various activities, such as pushing data through the network or even creating new actors. The internet in general is often a fire space, and producers of social web tools often consider being disruptive a metric of success. Certainly, times of disaster create fire spaces when everyday people and organizations try to react to events quickly. This term, moreover, sheds light on the need to reassess the ANT concept of general symmetry of agency among all actors. Each actor in the system has the agency to take action. Of course, the actions of technologies are often under human control—just because your alarm clock goes off does not mean you have to get out of bed. However, seeing each actor that is present in the system does allow us to understand the context in which people may act within the network. By understanding that agency can extend to both people and tools, architects can focus on the experiences of participants in a fuller, more holistic fashion. Yet, in fire spaces, agency does not necessarily disperse evenly across the network. A hashtag is only as effective as the participants who use it.

We can view such agency as ebbing and flowing across the network and from actor to actor, depending on the situation to which the network responds. This ebb and flow is yet another indicator of the persistent challenges to network participants as they scour the social web, news sites, information sites from emergency response agencies, and many other spaces for content. Across these locations, participants take part in a collective effort to write the most complete narrative. The strength of ANT is its ability to identify an assemblage of participants and, through careful analysis, describe "what they do to expand, to relate, to compare, to organize" the data they find (Latour 2005, 150). The barriers that separate these spaces place a major burden on participants who use them. The difficulty lies in finding not just data, but data relevant to the task at hand. Such a burden forces participants to expend enormous amounts of time and energy in addition to developing a sophisticated knowledge of the data structures of websites—websites that do not speak to one another. These participants must also gain a strong understanding of the functionality of search engines both for the web in general and for the specific websites they are using.

To effectively research and architect communication systems, we as researchers and practitioners must become active participants in such systems. Only through involvement can we uncover such detail. Through participation, we can better grasp the motivations that drive participants to perform these tasks. Locating relevant data can quite often be a source of frustration for network participants and researchers alike. Working as active participants can help us understand the requirements of future systems that collect and aggregate information in support of such efforts. We can also work toward developing systems that present information effectively and architecting tools that support this kind of data collection.

Agency of People and Technology

As mentioned earlier, ANT positions technology and people as equal agents of action and refers to both as actors within networks. These networks bring together individual actors to form assemblages, or series of networked relationships that might form for short periods of time or in response to specific events or needs such as helping to locate a missing person. In sum, ANT describes a network of human and nonhuman actors that assemble to push information among actors in an effort to find, validate, and share information.

As Callon (1986), Latour ([1996] 2002), Law (1992), and Mol and Law (1994) described in more detail, ANT uses the term "actor" to describe humans and technologies. This term can also refer to organizations and events that are part of these networks. This approach can be especially useful when examining issues that extend across multiple global systems such as the social web. Viewing these actors equally can help practitioners consider whether their processes and systems are useful for actual, as well as intended, audiences.

In practice, we must not exaggerate the agency of nonhuman actors. For example, a realistic assumption is that people have more agency than their cell phones. By examining issues through this lens, communicators can focus more closely on the interactions taking place within these user experiences. For example, they can focus on the humans using cell phones to interact with other humans instead of focusing on just the phones as the centers of interaction. At the same time, mapping the various nonhuman actors helps contextualize the conditions under which these people-powered experiences take place. Considering technologies as important members of the network allows practitioners and researchers to build solutions that take such tools into account. This merged view of technology and user is at the center of the ANT approach to interaction mapping, and the move toward using ANT to research issues in technical communication follows the recent innovations by scholars looking at similar issues of technologies, networks, and participants (McNely 2009; Potts 2008; Potts 2009d; Potts and Jones 2011; Spinuzzi 2008; Swarts 2009).

To understand usability issues in the social web, architects must find the key actors within these networks, for these actors are often in the midst of some form of relevant action, such as locating data, passing information through the network, or sharing knowledge with the community. Researchers can then trace this activity as it moves throughout the network. For example, by observing how participants consume digital content (such as how they download music to their iPhones), communicators can learn which technologies are useful (iPhones over laptops), which people are directing the participant activities (owners of iPhones), and which organizations and participants are involved (Apple, artists, fans, record labels, the providers of network access). Without tracing these actors and pinpointing their activities, researchers must rely on metrics showing that these actors came and went, rather than an understanding of how these actors made associations and how they looked across an entire ecosystem of technologies, people, and organizations.

ANT is a useful tool for examining the ways people, organizations, technologies, and processes can assemble into networks, make new connections, and work collectively. Combining ANT with more traditional user-centered design methodologies such as needs analysis and contextual inquiry helps us research these systems. Doing so also gives practitioners a more detailed way to trace these scenarios. With such knowledge, experience architects are better prepared to aid in the architecting of software, processes, and policies to support realistic, user-centered experiences. Diagramming the people, places, organizations, events, and technologies can empower development teams to know their audience's context, relationships, and distribution before they attempt to create innovations. These diagrams can help teams reach common ground more quickly as the teams share the visualization and discuss the implications of their proposed product designs, policies, and services.

Locating Data and Sources after Hurricane Katrina

By the time Hurricane Katrina was over, the disaster had affected millions of people across the region. Despite publicized predictions of when the storm would make landfall and how much damage it might cause, many people—including people in the federal and local governments—were unprepared for the level of destruction that followed. While my research focuses on the activities of everyday people in relation to sociotechnical use, research on disasters tends to focus on the actions of the media (Blakemore and Longhorn 2001; Rappoport and Alleman 2003) and everyday people (Carey 2003; Collins 2004; Vengerfeldt 2003). Many researchers have studied Hurricane Katrina (e.g., C. Jones and Mitnick [2006]), but at least one (Perlmutter 2006) emphasized how difficult writing about this natural disaster can be for academics in the United States because of how recent and personal the disaster was for us. As for aid workers, a former Red Cross worker noted that when phone communications are unavailable, people are likely to use the web, stating that "the internet can be a powerful element in crisis situations" (Putnam 2002). These articles point to the internet as a site of communication and coordination during times of disaster.

Though the social web has grown exponentially in recent years, sites for news agencies, government agencies, and nongovernmental organizations (NGOs) are typically still closed systems in that they do not support the exchange of information beyond their boundaries. For social web participants searching out information in the wake of a disaster, these design limitations are the source of many frustrations. By helping participants to locate and validate information during a disaster, experience architects can trace how participants build narratives across multiple systems. Understanding the complexity of these situations will inform the creation of more flexible systems in which everyday participants can exchange information at critical moments. Participants in such systems engage in networked communication; they forage for information and then assemble that information in an ad-hoc, but still coordinated, manner. Rarely sticking to one website, participants actively move among sites, gathering information and turning that information into knowledge as they share it with others. Presenting this kind of literate activity is something new, requiring a different lens through which to study these experiences.

Data that cannot transform into information can never become knowledge. Such data are stuck in a state of invalidation. Examples of validation include confirming the name of a missing person and confirming that person's status (dead, injured, safe, etc.). Misinformation is a common problem across the internet and, really, in everyday life. For data to become information, we must validate the data. For participants to validate the data, they must trace the data across systems. Once they are able to validate the data, participants then must be able to interact within these systems to tag the data as information. Thus, the content can begin to move into the next content phase: knowledge. But without some way to interact with other actors or the data, some space to post, or some place to connect, actors cannot make these connections visible. If actors do not make these connections visible, then the data cannot move forward. While participants willingly come together, identify problems, take on group roles, and then stabilize needs and definitions, they do not always form networks stable enough to perform data validation. One reason is the closed structure of the many websites they encounter. A relatively small number of people (as compared with the vast numbers that can be mustered through social media) often maintain these websites. The closed borders of such sites can overburden their owners as they try to manage the deluge of information, making efficient and timely updates of the sites nearly impossible. Without these updates, the sites are useless to those who need them most.

Yet, as I repeatedly demonstrate in this book, everyday people not only take on these tasks, but they also often go about the business of finding, organizing, distributing, and verifying information in ways that are both systematic and effective. The examples in the following section trace both the successes and the limitations participants face in the fire spaces they inhabit.

CNN's Safe List

After Hurricane Katrina, the websites for mainstream news organizations such as CNN proved completely inadequate in attempts to collect and distribute information to hurricane victims and their families. Even though CNN (2005) created the Safe List to keep people informed about those missing and found, the organization's efforts were symptomatic of a top-down approach entirely in opposition to the organic activity that participants could accomplish in the social web. Those trying to work with CNN's Safe List found only frustration due to poor interface design and a walled garden of information enclosing information behind barriers. To change any of the information on the site, visitors had to email CNN's chosen moderators, who then would change the information manually. Instead of inviting the larger community to become actively involved in gathering, sorting, and validating data about hurricane victims, CNN did not trust the community to perform this sort of work. As a result, the Safe List could not evolve dynamically through the work of network participants.

The CNN Safe List website attempted to catalog hurricane victims who were safe, but the site created a frustrating and exhausting experience for those seeking information. It also constructed barriers for those participants who had information to share. Although CNN provided no way to edit the content of the Safe List from the website, the organization did provide a link to email the Hurricane Victims Desk, where people could contact CNN about the list. The design of the list itself did not make it easy to locate missing persons, and it did not provide any interactive features beyond moving from page to page. In that way, CNN's Safe List became a black box, or at least gave the appearance of becoming one in relation to ownership and agency. This section of the website is no longer active, but figure 3.1 illustrates its original information design problem.

Although the Safe List listed names alphabetically by last name, the list was more than 50 pages long with no indication of which letter started on which page. To find the name "Bill Lewis," participants had to guess—would the Ls start on page 20? Page 30? A numbered list is not an intuitive implementation of an alphabetical data set. Where this method might be merely inconvenient in everyday searches of address books or playlists, we can only imagine the mounting frustration and panic as the relatives of a New Orleans resident hunted, via trial and error, to find one name among hundreds in this system.

In visualizing how certain data were central to this event, we can map out the actors in this network. Within this example, the data could not move into information through the CNN website because everyday people were unable to participate in a transparent way. Therefore, this ANT diagram may look incomplete. I did this on purpose to illustrate the lack of participation in this network. In figure 3.2, we map this actor network, putting the Safe List in the center. Around this central artifact are all of the associated actors that must be present for this central artifact to exist. To have the Safe List, Hurricane Katrina must have occurred. Without this event, none of these actors would be present. CNN, and in particular the Hurricane Victims Desk, is obviously a major actor in this network. Another actor relevant to this map is Neighborhood America (now INgage Networks), named as a supporter on the website. Further tracing shows that this company builds social systems for other businesses. Its work in this network is critical to the space and gatekeeping design of this example. Also

Hurricane Katrina: Safe List

In Katrinas path: Reported safe

CNN.com has been posting the names of those who wish to let loved ones know they are all right after the storm. If you were in Katrinas path and want to post your name here, please send an e-mail to the Hurricane Victims Desk. For each person you are reporting for the list, include first and last name, age, hometown, state and a brief message. You may also include a phone number or e-mail address where those on the list may be reached. The list will be updated regularly.

Supported by Neighborhood America ▸

Page 1 2 3 4 5 6 7 8 9 10 11 12 13 14 15 16 17 18 19 20 21 22 23 24 25 26 27 28 29 30 31 32 33 34 35 36 37 38 39 40 41 42 43 44 45 46 47 48 49 50 51 52 53 54 55 56

LAST NAME	FIRST NAME	CITY	STATE	MESSAGE
Acosta	Scott	New Orleans	LA	n/a
Acosta	Ray Jr.	n/a	n/a	Safe in Houston, TX
Acosta	Orelia	n/a	n/a	Safe in Houston, TX
Adams	Angela	New Orleans	LA	Evacuated first to Texarkana, but is now in Lake Charles, LA
Alecha	Francisco	New Orleans	LA	Evacuated to Houston, TX
Alecha	Enriqueta	New Orleans	LA	Evacuated to Houston, TX
Alegria	Janice, Angel and Melody	Chalmette	LA	Evacuated to Houston, TX
Alexander	Les	Metairie	LA	Evacuated to Orlando, FL
Alfonso	Jeannette and Leroy	St. Bernard Parish	LA	Evacuated to Concordia Parish Community Center, Ferriday, LA
Allain	Donn, Jeanie and family	Biloxi	MS	Safe at home in Biloxi, MS
Allange Jr.	Mike and Lindsay Manguno	St. Tammany Parish	LA	Safe in Marietta, GA
Allen	Elijah (Fuzzy)	n/a	n/a	Evacuated to Atlanta, GA
Allen	Shyra and Devonte	n/a	n/a	Evacuated to Austell, GA
Alpert	Henry	New Orleans	LA	Safe in Baton Rouge, LA
Ambler	Rosalie	New Orleans	LA	Evacuated to Mobile, AL
Ambroza	Abe	New Orleans	LA	Evacuated to Lafayette, LA
Amole	Vicki	New Orleans	LA	n/a
Anderson	Elizabeth	Diamondhead	MS	Evacuated to North Carolina
Anderson	Dr. Milton	New Orleans	LA	Safe in San Antonio, TX
Anderson	Johnna	n/a	n/a	Safe in Shreveport, LA

Page 1 2 3 4 5 6 7 8 9 10 11 12 13 14 15 16 17 18 19 20 21 22 23 24 25 26 27 28 29 30 31 32 33 34 35 36 37 38 39 40 41 42 43 44 45 46 47 48 49 50 51 52 53 54 55 56

International Edition   [Languages ▾]   CNN TV   CNN International   Headline News   Transcripts   Advertise with Us   About Us

SEARCH ◉ THE WEB ○ CNN.com [   ]   (SEARCH)   Powered by YAHOO! search

FIGURE 3.1 CNN's Safe List of the Hurricane Katrina missing. The website is no longer online.

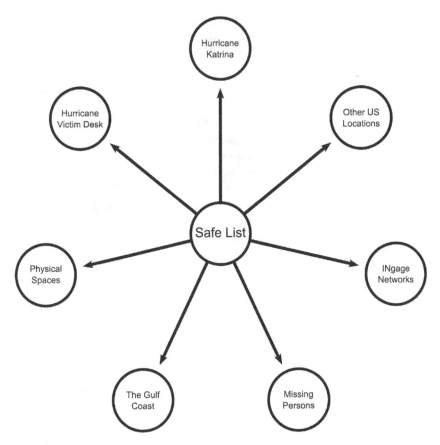

FIGURE 3.2 Actor-network diagram of the CNN Safe List.

vital to this network are the missing persons that the Safe List attempts to locate. Included in this map is the primary area that this disaster affected (the Gulf Coast), as well as other locations in the United States where friends, relatives, and evacuees might have relocated. Equally vital are the physical spaces where these victims may have relocated.

I have included in this map the participants who are never quite part of the interface. Such people include victims, evacuees, friends, relatives, and any other members of the community who want to add, delete, or modify the data in this system. Note that these participants are not connected to the map only because their ability to participate is not evident in the interface; although they can presumably participate by emailing the Hurricane Victims Desk, the details of participation are unclear regarding what this process might entail, how gatekeepers validate content, and who decides what to include. The content remains immutable, stabilized within the black box of CNN's Safe List. This content is data and does not undergo community validation. It is immutable, unchangeable on the fly by participants. In this map, I document this disconnect in a literal way for a number of reasons. ANT diagrams can help visualize the context in which these experiences are occurring. Having these maps helps teams understand the people, technologies, and organizations that any changes may affect. Such changes could be new technologies or even new processes.

Again, those trying to work with CNN's Safe List found roadblocks due to poor interface design. This walled garden lists only data, hiding information behind the interface. How do we know that any of this content has been verified? And if it has, by whom? We have no source, no links to confirmations. To change any of the information on the site, visitors had to email CNN's chosen moderators, who then would change the information manually. Instead of inviting the larger community to become actively involved in gathering, sorting, and validating data about hurricane victims, CNN did not trust the community to perform this sort of work. As a result, the Safe List could not evolve dynamically through the work of network participants.

In understanding who is involved in this network, who might have agency, who might have frameworks supporting their work, and who might be part of the crowd, we can create stencils to illustrate the kinds of actors present in this network. Selecting unique stencils for the diagram can identify the different actor types by showing commonality in the stencil type (see figure 3.3). For example, we might represent individuals with an outline, or silhouette, of one person, and we might represent a group with an outline of multiple people. We can represent an event such as a hurricane with a symbol of that kind of event. If we wanted to represent multiple disasters on a single map, we might create one universal symbol for disasters. Teams of researchers and architects should base their stencil choices on what their team wants to emphasize.

The next stage in mapping this actor network is to define the relationships among the actors in regard to content (see figure 3.4). Depending on the focus of the research and discovery, the relationships mapped in these diagrams could be vastly different. Examples include showing closeness of ties among actors, showing how often actors connect with each other, and showing which actors have permission to share content with other actors. Not every ANT diagram needs to show relationships, but drawing these lines to illustrate some sort of connection among actors can be useful.

In this case study, participants were unable to interact with the system to add important details, edit names, locate duplicates, or point out incorrect entries. This lack of interaction significantly stymied any ability of this network to reach translation (i.e., for the network to form), much less work through the different stages of this process. Imagine the frustration of locating a lost child or missing person and not being able to make that information public on the CNN site to reunite relatives. Those in a position to help, as well as those who needed information, were completely reliant on what CNN decided to release to them from

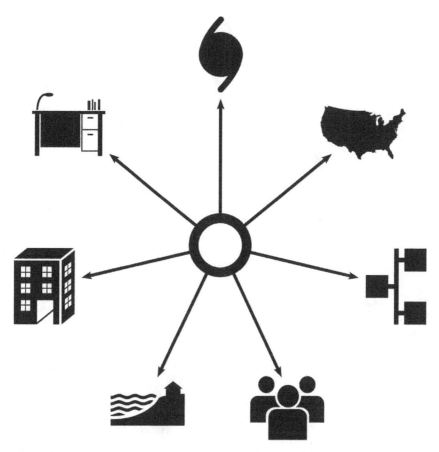

FIGURE 3.3 Actor-network diagram with stencils of the CNN Safe List.

within the virtual walls of its news outlet. Therefore, in this diagram, some of the actors are not connected to the central actor. The strongest ties are between the event itself and some of the actors. The weaker ties are between the less critical actors and the more important actors. In this case, the emphasis of ties is relevant to how the central actor moved through the network. The goal of this map is to try to help visualize the context in which these experiences take place.

On static websites like CNN's, participants have "no trace left, thus no information, thus no description, then no talk" (Latour 2005, 150). Actors would have benefited from knowing where the site authors were finding their content, how they were verifying this content, and where site visitors could go to dispute this content. All of these key activities would have helped to validate the content. Without these traces of information, traces of actual human contact were lacking. The supporters who lend strength and credibility to the public discourse, the "mass of silent others" (Callon 1987, 90), lacked an entry point from which they

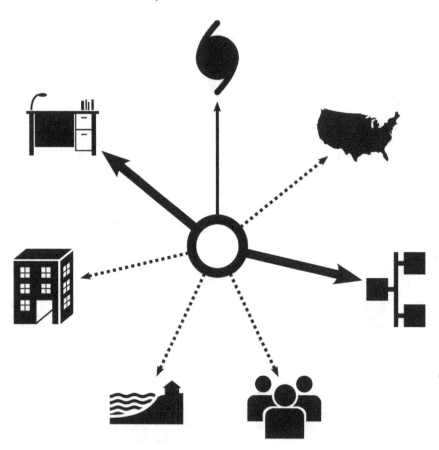

FIGURE 3.4 Actor-network diagram showing relationships among actors in the CNN Safe List.

could participate in validating and updating information. Whether by intent or oversight, CNN became an information gatekeeper by not allowing participants to add to or modify the content of the Safe List.

In other cases during this disaster, participants encountered websites that buried much-needed information within the website instead of making it easily accessible from the home page. In response to Hurricane Katrina and family members looking for lost or missing loved ones, FEMA's website initially directed visitors to the Red Cross website designed for the purpose of locating missing persons. And even though creating a simple hyperlink can be an immense help to participants who are searching for loved ones, this simple option can lead to several problems. Clicking the link first took participants to a warning page stating that they were leaving FEMA's website and that FEMA was not responsible for the information on the other site. Once the participants arrived at the link's destination, they found themselves on the Red Cross home page instead of the specific page for finding missing and displaced persons. If visiting the FEMA website was the interactive equivalent of asking, "How can I find my loved one?" then FEMA's link was the equivalent of shrugging shoulders and replying, "How should we know? This experience not only actively hindered the ability to make connections across the web, it undermined a sense of trust or support that visitors to these websites needed. Instead, visitors found themselves essentially left on another agency's doorstep. This example illustrates how mobility can go wrong with an absence of understanding cultural use and supporting meaningful content.

In the New Orleans section of the classified-advertisement website Craigslist, participants appropriated the lost and found section and turned it into a mechanism for sharing information on post–Hurricane Katrina New Orleans. People posted requests for information about friends and family who had been trapped in New Orleans during the hurricane or haphazardly evacuated to other regions of the United States. Refugees and victims also posted information in attempts to notify their family and friends of their health and well-being. In other cases, Craigslist participants posted housing offers for victims and refugees, though not always with the best intentions.[2] While the use of Craigslist as a site for activity after a disaster was unexpected, the location-specific sections of Craigslist proved useful for quickly posting content. People made many attempts on Craigslist to contact missing individuals. Unfortunately, many websites allowed participants to locate and reconnect with others, but these sites were unconnected. These different websites led to multiple postings across these sites with no method of tying together all of the information and to a total disconnect between the lost and found. Undeterred, hundreds of Craigslist participants posted offers of shelter.

These examples demonstrate barriers that block information validation, leaving actors with only data and no easy way to move content across the network, a situation antithetical to the goals of the social web. Instead of supporting an actor network that spans the web and enhances the capabilities of actors across that entire network, these systems close off these interactions, segregating their information from the social web and keeping the information invisible to the wider range of people, technologies, and tools that can readily use it. In doing so, these systems make the processes of validation in the enrollment stage of translation far more difficult. Within the websites themselves, these processes are nearly impossible. Participants cannot create usable traces from these sites. Thus, translation cannot occur, and validation cannot take place, leaving this content in a constant state of existing only as data.

Practical Solutions

Looking specifically at the examples I have examined in this chapter, locating missing persons is paramount in the wake of a disaster. The simple act of tracking down data that friends and relatives can validate is often mired by inaccuracies such as misspellings, location mismatches, and even mischief. Participants must be part of these data-gathering activities so that they can also help to later validate that content and redistribute it to the community as knowledge. Such issues of misspellings, mismatched data, and mischief not only occur with disasters; these issues occur throughout the internet just as they occur in daily off-line life. Sometimes these inaccuracies are issues of purposeful gatekeeping, sometimes they are due to a lack of understanding of how to build for participation, and sometimes these barriers exist because the content producers simply do not realize that participants want to do certain kinds of activities in these spaces.

Often, implementing solutions depends less on technology and more on the policies of the people managing these systems. We have the technology now to build alphabetical lists, to allow for the sorting of columns in tables, and to create other simple affordances. We can now build location-specific solutions, deploying mash-ups that place content onto maps so that people can visualize geographic data. We can create interfaces that allow participants to add, edit, and delete content. Allowing everyday people to be part of the solution to data problems also allows for more chances for validating content into information. It allows participants to coproduce content, to feel they are part of the solution. For organizations, having participants become active members of their sites allows the site owners to leverage this knowledge for social good—the kind that can lead to awareness as well as give them the right kind of publicity.

Much of the interaction of participants on Craigslist aimed at finding missing individuals through posting messages in the lost and found category of the New Orleans section of the site. Ordinarily, Craigslist participants post in this category information about pets, jewelry, or other items that they have either lost or found, hoping to reunite owners with their belongings. However, participants quickly and heavily appropriated this category by discussing the disaster, in some cases with offers of jobs or housing and in other cases with requests from people trying to locate missing friends, loved ones, or pets. Participants transformed the category as they appropriated the prescriptions and conventions of the Craigslist community by reinscribing it as a missing-persons list.

We can build social web tools and other systems that need participation to allow data to flow and participants to become active members. For those system owners who are concerned about content integrity and ownership, we can also now build systems that provide multilevel permissions based on whatever business rules that the organization might be following. Trusting moderators to help curate content is a useful tactic, as is having some sort of trash collecting for irrelevant data. After the Haiti earthquake of 2010, Google's people finder (now defunct) was inundated with fake information, and it desperately needed moderators to comb through the content and clean it up. Moderators can do this kind of work if they understand the culture they are working in and have the tools to curate the content. But they need to be welcomed in to the system—something I examine in the next chapter.

In the aftermath of disasters, people need to be able to turn data into information by validating and adding details to it. Participant-generated, ad-hoc composition is becoming an increasingly common manner of production, as recent researchers have noted (Johnson-Eilola 2005; Slattery 2007; Spinuzzi 2003, 2007; Swarts 2007, 2008). Experience architects need to build for this interactivity. Mapping these networks can help development teams visualize the context in which these interactions take place. In the case of the CNN Safe List, days passed before the site's producers could refine this format into a more user-friendly list. News outlets such as CNN, organizations such as the Red Cross, and government agencies all failed to create flexible, open systems that would have allowed participants to add and edit content and developers to push and pull that information from the databases that held it. Such spaces would have allowed participants to help validate information, track the missing, and triangulate data. Allowing everyday people to participate in these spaces can lead to richer information with volunteers aiding in the curating of these spaces.

Conclusion

Participatory culture can only flourish when we welcome and support participation. By empowering volunteers through spaces that allow for interaction, we can harness the power of the crowd to aid in distributing data and validating this content. In this chapter, I discussed the limitations that participants faced when they worked to validate information after Hurricane Katrina. As we identify the needs of the actor network and define the terms of its operation, contextualizing the data that network participants find is of crucial importance. We need to cross-reference this data and continuously update the data in a timely and effective way. Because the systems that I discussed in this chapter could not support this transformation from data to information, the ability for data to move forward into the state of knowledge that participants could then repurpose and redistribute to the community was impossible.

The attempts of CNN, FEMA, and the Red Cross were certainly noble, and we should not see them in a negative light because of their inadequacies. Mainstream content creators wanted to provide solutions for connecting actors in order to locate missing persons. However, the level of validation (and lack thereof on Google and CNN) created frustrating experiences that did not allow for translation to move forward. Craigslist was a success story coming out of this disaster, as it connected people, helped them locate the missing, and encouraged community (Axline 2005). We must work to build systems that can allow for this level of participation, but in a way that can organize content, help people locate the data they need, and allow these people to validate this content for others. As I noted earlier, the lack of access for many victims initially hindered their use of Craigslist, just as it hindered the use of sites by CNN, FEMA, and the Red Cross.

Yet, the Craigslist site's flexibility as a social tool allowed participants to step outside of its intended use and repurpose its design for their own needs. Making this space mutable was key to creating a participatory structure. Craigslist's openness is a feat that larger sites behind gated barriers—barriers that a small number of administrators or owners maintain—have not managed to accomplish.

The examples in this chapter point to two major extensions of and challenges to the theory that I discussed at the beginning of this chapter. First, and most obviously, agency is not equal. While we may have the technologies to solve some of the problems that I have discussed, we do not yet have the agency to make these solutions happen. While the CNN website's structure looked as if it could invite discussion and validation of content, participants did not have the agency to scale the Safe List's walled garden and engage with the content and content producers. The examples that I discuss in this book's final chapter illustrate that even when agency is equal, we do not necessarily have a better solution. In the case of the Haiti earthquake of 2010, when Google created a missing-persons list that anyone could edit, participants ended up getting fake data. In the case of the more recent Hurricane Sandy in 2012, equal access to Twitter among participants did not end misinformation. The next chapter explores the role of anchor actors, or volunteer participants, who can help parse this data and aid in the validation of content.

As the second challenge to the initial theory, participatory culture questions the concepts of mutable and immutable mobiles. The idea of a single, solid, unchangeable mobile is an easy concept to hold on to in spaces that are stable and predictable. Fire spaces race to create these immutable mobiles, creating order out of chaos. However, a single entity does not always win this race. While CNN had a Safe List, Craigslist had its own list that was powered by everyday people who were desperate to connect in a space that would allow them the agency to do so. We need to consider solutions that allow for both kinds of activity, letting the official and participatory streams flow together, overlap, and provide the agency to move content from data to information. The next chapter examines how this activity took place in more unofficial spaces where more participation was welcome.

The solutions experience architects create for future systems must take into account each stage of translation. Participation within the social web continues to increase exponentially, and the number of social networking systems increases as well. This evolution offers opportunities for architects to provide solutions that empower network actors to be active participants. Communicators can provide tools that connect social networks with mainstream media sites that hold data; communicators can then empower anchor actors to direct participants through stages of translation (formation of the network) more effectively.

4

VALIDATING INFORMATION DURING
THE LONDON BOMBINGS

On the morning of 7 July 2005, London's public transport system was the site of multiple coordinated attacks by suicide bombers. In what is known as the 7/7 bombings, 52 victims were killed and more than 780 were injured (Home Office 2006). Targeting the London Underground subway system and a double-decker bus, these coordinated attacks occurred at four locations across London. Around 8:30 a.m., three bombs exploded on subway trains near Edgware Road station, near Liverpool Street station, and between King's Cross and Russell Square (*BBC News* 2012). Occurring during the morning rush hour, these attacks took place as many of the victims were on their way to work, carrying various technologies such as cell phones and laptops. With these devices, victims were able to communicate with friends, relatives, organizations, and strangers as they tried to piece together what was happening.

In the aftermath of disasters (both natural and human-made), people communicate by cobbling together available internet resources, relying on the capabilities of various social web tools. The examples in this chapter illustrate how participants in the social web work to verify and validate information across networks. Information is validated data: text, images, videos, audio, and so on. After they validate information, actors can repurpose this content, sending it through the network as knowledge. These examples illustrate how participants turn data into information, repurposing that data and linking people, websites, social networks, and other sources and technologies. Observing and analyzing how individuals and technologies verify information are central to understanding how architects can better research these interactions and architect for these experiences. By looking specifically at high-pressure cases such as disasters, researchers can see how participants use the existing tools, and experience architects can develop new solutions for products and services.

In this chapter, I map out research methods and analyze how participants validate source material during their search for answers and information related to crisis events. Pinpointing instances of interactions where participants cross-reference information is critical to architecting and implementing more effective communication networks. As experience architects, we must examine how participants evaluate data, navigate poorly designed systems, and connect with one another through these networks. The case in this chapter further illustrates the utility of this approach as a method for understanding how participants in the social web verify and validate the data they collect, thus turning the data into useful information that can help others in desperate need.

The case that I examine here shows that flexibility in social web tools is necessary to encourage participation in these spaces. For people to validate content and communicate more effectively, they need tools that provide support for these activities. During crises and disasters, information is critical for all those involved. It can help rescue victims, aid rescue and relief workers so they can do their jobs more effectively, help identify victims, and provide a much-needed catharsis to friends and loved ones. Participating in these actor networks also provides a sense of comfort, belonging, or even simply emotional relief to many of the participants—even those not directly affected by the events—just as it does in everyday experiences for knowledge workers, colleagues, and Facebook friends. Researchers and architects can use these methods and examples to explore how to better understand and improve their tools for daily use. The research methods and examples that I outline here lead us toward solutions that help participants perform the tasks of data triangulation and validation.

Ecosystems and Information

For the purposes of this book, I define "information" as content validated by participants in the social web. Validation occurs when an actor shows evidence that the content is proven true for a given culture and community (e.g., connecting the image of a missing bombing victim with an image of the found victim in the hospital). During disaster events, images of missing persons propagate across websites such as those of the Red Cross and *BBC News,* Facebook, and Flickr. Rather quickly, missing-persons pages on Facebook appear. Sometimes, people take photos of missing-persons posters and upload the images to Flickr. Then, we begin to hear stories of people either deceased or hospitalized. Volunteers on the social web attempt to connect these two different kinds of content: official and unofficial. In doing so, they are validating the missing-persons posters by linking them to the articles about the missing persons. They verify data as an attempt to help friends and relatives of the missing persons and to find their own closure regarding individual stories of the event.

After participants have located data, they often focus on answering several questions in attempts to validate the content. For example, they may ask how reliable the source is, how accurate the information is, and whether any of it can be confirmed elsewhere (or denied). Sometimes, they can find these answers through official channels such as police agencies, emergency response teams, or relief organizations. But these organizations do not always provide information that is timely, findable, and usable. Organizations such as the Red Cross and local municipalities are only just starting to understand and implement solutions for rapid internet response in emergencies.

Other times, this information is buried on websites that search engines cannot locate, as is often the case with websites launched for a new event that have yet to build up links and connections that would help them surface. Many times, this information is in spaces that do not invite participation, where content producers have locked down and controlled content, as was the case with the CNN website's Safe List during Hurricane Katrina. To validate content, people need spaces where they can note these validations and link to references. How can participants be motivated to follow up on their interaction and further trace the information they find? We need them to be motivated so that they participate, whether this participation is to help track down a missing person or to access a relevant article. Social web tools need to be built to support such participation and to help participants answer these questions in timely, context-rich ways.

The substance of such participant-based questions revolves around the need to triangulate and retriangulate scattered and evolving bits of data in order to validate them as relevant, accurate information. Networks are only as strong as the connections among participants. Often, a group of participants rises up and tries to organize, validate, and distribute content. We can refer to these actors as "anchors," because they are trying to fasten the network to a set of content and content types, such as websites and images, that make sense of the event and form the network. The duration of these connections might sometimes be the scale by which researchers measure the effectiveness of the connections; however, more often the effectiveness is better measured by an anchor actor's ability to cross-reference information or let other actors leverage that information. Dumping massive reams of data into a network is meaningless if actors can do nothing with that content (Brown and Duguid 2000). The network must be usable. In other words, systems and tools must do more to help, rather than hinder, network participants as these they gather, distribute, and validate information. To develop more accommodating communication systems for everyday use, researchers and architects can examine participants' literate appropriations of tools and information.

Validating Information

The method I discuss in this chapter is based primarily upon actor-network theory (ANT). An actor is typically any human or nonhuman entity assembled into a network to perform an action. For the purposes of discussing this framework as a research method for understanding how to build systems, we can extend this definition to include any human, technology, organization, group, event, and artifact. In a work setting, these actors might include managers, products, salespeople, developers, and competitors. An actor network comprises actors who move content through the network and carry out various activities to support this movement. For the sake of understanding network effects, we tend to focus on the actors who are active participants in the network. ANT helps us examine these movements, systems, and participants so that we can assess the technologies, processes, and practices in context. Such moves can help us understand the validation process by which data becomes information.

We can trace this movement of content by using a framework that helps us focus on these actors and their activities. Callon (1986) defined these four moments of translation as problematization, interessement, enrollment, and mobilization. These moments can be fluid and can overlap. By locating activity in networks and examining moments of exchange among actors, we can trace these moments by observing when

- anchors and other key actors first recognize that an event is taking place and volunteer themselves to participate in the network (problematization);
- the identity of actors in these networks stabilizes and the allies align (interessement);
- actors accept the new system and coordinate action (enrollment); and
- the allies mobilized, the anchor actor is the spokesperson for the network, and punctualization is reached (mobilization).

Through the activities that transpire during these four moments of translation, a coherent network forms, and knowledge can pass through the network. By tracing these movements among actors, we can understand how translation happens in these ecosystems. Typically, the connections among actors become obvious through the efforts of anchors and participants. These actors work with immutable and mutable mobiles. As noted previously, immutable mobiles cannot be changed, but mutable mobiles can be. An example of an immutable mobile is a hashtag such as #electionnight. Before becoming immutable, that hashtag could have taken several forms, such as #electionnight2012, #election, or #electionnight12. Actors come together to push these mobiles through the network, sharing data, validating information, and repurposing knowledge. Here, I focus on the first two moments of translation (problematization and interessement) to trace and examine content transformation from data to information. In the next chapter, I shift focus to enrollment and mobilization to discuss how information becomes knowledge.

By using a portion of ANT's four moments of translation, we can pinpoint how data become information. The concept of translation is a central point for understanding how these actors do their work. During the process of translation, the group of actors reaches punctualization—the group becomes greater than the sum of its parts—as the actors work to create a centralized network. For example, on Flickr, individual photos are spread out all across the network. Photo pools allow people to collect and curate images across Flickr. People can add images to these photo pools based on interest, geographic location, event, or other topics. When people add an image to a photo pool, the image becomes part of that community and can then be shared with a wider audience within Flickr. A photo pool of missing persons is greater than the individual photo in that the pool is contextualized, easier to locate, and curated by an anchor actor. To walk through the activities involved in translation, we need to look at the multiple stages that take place during this process.

Researchers work to uncover these connections by focusing on the relationships among actors and by watching the ways mutable mobiles move throughout the network and the changes they absorb from actors. Mutable mobiles such as Flickr photos containing metadata are fluid artifacts that carry information throughout the fire space of the social web (Law and Mol 2003). Information that describes data is metadata. The metadata of a photograph can include time stamps, camera types, geographic location, and other information. Rapid connections (which can just as rapidly disappear) form within collaborative spaces, thus creating fire spaces where activity is kinetic. The movements of mutable mobiles are unpredictable, and because other actors can change them, mutable mobiles are not stable artifacts. As I discussed earlier, from their movement, mutable mobiles gain stability—not as static representations, but as vessels for moving information around the network. Mapping the network and tracing the ways participants validate information is a matter of following mutable mobiles through the fire space of the social web. By following them, researchers can locate points of articulation and analyze the changes participants make to mutable mobiles.

Using ANT, researchers can pinpoint and visualize the networks that various actors validating information create, during activities such as when victims are found alive, when news emerges, and when other details surface. By knowing who is participating, where they are participating, what systems are supporting them, and what they are doing, we can better understand how this translation happens. Tracing the interaction dynamics that support forming an information network is critical to understanding the context in which these activities occur. Such tracing can help architects of existing systems add additional tools that support these experiences and structure metadata so that the metadata can freely flow wherever during times of disaster.

Problematization: Emergence of Participants and Anchors

The first moment of translation is problematization. Callon (1986) defined this moment as the time when actors seek "to become indispensable to other actors in the drama by defining the nature and the problems of the latter and then suggesting that these would be resolved if the actors negotiated the 'obligatory passage point' " (196). During a disaster, volunteers who share content or organize that content usually perform the activity of assigning the obligatory passage point. Those indispensable actors are anchors. They perform these acts by uploading images to spaces such as Flickr, Tumblr, and Instagram, creating blogs dedicated to certain events in hopes of aggregating data and building groups on Facebook to organize volunteers and share information.

Anchors help organize and distribute this content either within one space or across an ecosystem. Each connection across the network provides a way for information to travel to new actors, become verified, and eventually be redistributed to the community as knowledge. Often, these obligatory passage points are volunteer members in a community. Sometimes, they are human volunteers eager to add to the conversation. Other times, they are technologies such as cell phones with embedded geolocation data that empower information validation. In the majority of these cases, obligatory passage points validate content by connecting disparate pieces of data and verifying them with evidence in images, news articles, blog comments, or other kinds of source material. Whoever or whatever has this role, the obligatory passage point holds power over which data transfer across the network, how participants validate that data, and when they redistribute the data as community knowledge. We often see volunteers eager to help, and in helping they can be empowered by each other and technology to help translate information across the network.

Interessement: Stabilizing Actor Identity

Callon (1986) described the second moment of translation as interessement, "the group of actions by which an entity . . . attempts to impose and stabilize the identity of the other actors it defines through its problematization" (208). At this point, the actors are encouraged to accept the anchor actor's definition of the event and key actors. Often, the anchor actor is a human organizer, moderator, volunteer, or even a technology, alternating roles as the pathway for the obligatory passage point. Defining terms and stabilizing language, these anchors organize content by validating the relevancy of this material. Sometimes this validation occurs when hashtags stabilize on Twitter, meaning that one hashtag emerges as the dominant hashtag that people use to describe an event such as a tornado or an earthquake.

An anchor's ability and eagerness to aggregate content and encourage others to add to this content pool are the first steps toward transforming a community into a network. For such translation to take place, the participants must have the system literacy to use these sites. The need for system literacy may be one reason why these networks often center on familiar locations where participants already have experience using the systems and why participants do not turn to new sites invented specifically for a particular disaster or emergency. This familiarity is a base from which anchors can bring in content aggregated from other spaces to confirm various claims made by the community. By stabilizing identities ("I am the anchor; you are the helpers; this is the site we are using"), information can begin to take shape based on these commonalities.

Tracing the Translation from Data to Information

The case that follows illustrates the utility of using these first two stages of translation to examine how data become information. For researchers using ANT, this method allows them to highlight different ways in which network participants assemble. Doing so can allow them to then examine system performance in regard to network participation. Such an examination allows us, as researchers and practitioners, to ask key questions. How well does the system support participation? Can our technologies better serve our participants? How can we make improvements? By tracing these activities, researchers can locate the ways in which actors assemble. These actors must work with each other, forming the network and mobilizing to reach their goals. Though this case study follows the general pattern of the ANT method, I describe each instance to highlight different stages of this method and the usefulness of the stages for researchers and experience architects. This chapter looks at how the first two stages of translation played out during the London bombings. The next chapter looks at the final two stages as they unfolded during the Mumbai attacks. Applying this method to several different examples demonstrates ANT's adaptability and effectiveness for researching separate networks that focus on different types of artifacts, seek different sorts of knowledge, or work with different web spaces and social web tools.

Figure 4.1 illustrates many of the technological actors leveraged on 7 July 2005 to find and exchange information. As the diagram shows, many different technologies were involved in this network, including blogs, wikis, and media sharing. Also illustrated are a number of different types of posts that participants made in a variety of ways to distribute updates, news, and other information. However, figure 4.1 also illustrates some of the challenges that participants faced as they moved from one site to another in an attempt to locate information and develop methods for distributing it to other actors. As stated before, many of the websites and tools that the participants used in this case were functionally separate from others. The effect was akin to creating walled gardens that prohibit the cross-pollination of information. The moves to form the actor network fell to the conscious, directed actions of human actors as they sought to find work-arounds for overcoming the barriers among technologies. In other words, the segregation of technological actors hindered the information space that actors wanted to create. Systems did not talk to each other easily, if at all. Finding, aggregating, and distributing information around the network thus meant working across the spaces shown in figure 4.1.

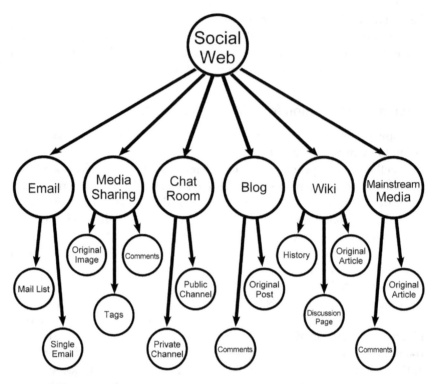

FIGURE 4.1 Systems of activity.

Essential parts of this activity are the anchor actor and other volunteers to help with collecting data, linking information, and posting content. Equally important are technological infrastructures in place such as websites and text messaging systems. The anchor and the volunteers can be various actors—technologies or human beings—such as RSS (Really Simple Syndication) feeds, forum moderators, search engines, and blog posters. Together, this group forms the beginning of a network that can be made whole and moved forward by a spokesperson who can lead in the translation process.

Through this application of ANT, researchers can visualize the networks that the various participants create. Knowing who is participating, where they are participating, which systems are supporting them, and which technologies are in use helps inform the development of systems that can support these activities. This kind of knowledge can lead to new features and support structures for allowing content to move across these ecosystems during times of disaster and peace. Researchers must trace anchors as they locate information across multiple systems, collect this information, and work with the community to validate it. In this scenario, technology and policy could support the participants' acts of composition and coordination.

In our examination of this case study, understanding the context in which network actors were engaged is important. When the London bombings occurred, participants had an extremely wide range of technologies that helped them to gather and share information almost immediately after the attacks took place. In 2005, the idea of accessing the internet through a mobile device such as a PDA or cell phone was already well established. And London commuters quickly turned to these devices to post updates to blogs and social networking sites—updates that often included photos taken by the onboard digital cameras common to most cell phones. One example of such use is the image of Adam Stacey taken as he escaped the Underground (see figure 4.2).

Across multiple websites, this image of Stacey escaping a subway car was the central artifact that helped participants define the network, allowed a space for moderation, and helped stabilize the identity of actors. This example helps us trace how data (the image of Stacey) became information (a terrorist attack occurred—we can validate the image as part of this event). This section describes this work in detail, showing how ANT can help trace these activities so we can understand how to do this kind of research and enable practitioners to build systems that can support this work. As I discussed in chapter 3, CNN's Safe List

FIGURE 4.2 Image taken of Adam Stacey during his escape from the London Underground, posted by Alfie Dennen (2005). Image by Adam Stacey, available under a Creative Commons 2.5 license at http://moblog.net/view/77571/.

did not provide a space for participants to engage with the content; in contrast, sites such as blogs and Flickr do allow for some level of engagement. For actors to validate the data they find in these networks, systems must allow them to participate in these networks. This case illustrates how validation can occur, although in this case, it stops short of reaching community knowledge. I discuss that topic in the next chapter as I examine a tipping point for participation on the social web.

Defining the Network

At the beginning of an event, data are abundant and information is scarce. Photos, tweets, and wiki posts are only beginning to emerge. Not everyone knows that something has happened, and misinformation can and does happen. As participants gather data, some of them begin to take on leadership roles. Certain actors become more relevant than others, becoming anchors that help form and stabilize a network. Here is one point where Latour's concept of equal agency and symmetry in ANT fails in reality to deliver on its promise (1999b). Some actors have more agency than others; some actors are more critical to the network than others. Without agency, a network cannot form. Anchors are key to these formations. What we want to pay attention to here is how anchors become useful and how we, as researchers and practitioners, can build for these networks so they can come together as needed. Supporting these activities is key to creating relevant and context-rich social web tools.

Such activities can take various forms. For example, on blogs, authors can tag their posts as a way to describe the content of the post. On Flickr, participants can tag images. In both cases, if given permission to do so, readers can also tag this content. Specifically with Flickr, participants wanting to respond to the London bombings blasts first had to find each other by accessing related tags through search or by clicking on an image tag, which would link to all other content tagged with that same word or phrase. Eventually, as an event progresses, Flickr participants can also find each other through photo pools, which I discuss later. In response to the London bombings, the largest such photo-pool group called itself the "London Bomb Blasts Community," and the tags found in this pool are seen in figure 4.3.

Many of the early photos uploaded to Flickr about the London bombings made their way to the London Bomb Blasts Community because of the work of the group's anchor, David Storey, who used the Flickr username "fgt" (Storey 2005a).[1] The group in this example comprises the anchor, Storey, and the various actors—cell phones, Flickr, links, photos, participants—that created the narratives sustaining this experience. However, for this network to come together, these actors had to first define what was occurring and then begin to negotiate terms for defining what they were encountering. Locating images and piecing together tweets and status updates, anchors and participants work to understand what is happening, where it is happening, and whom it is affecting. In doing

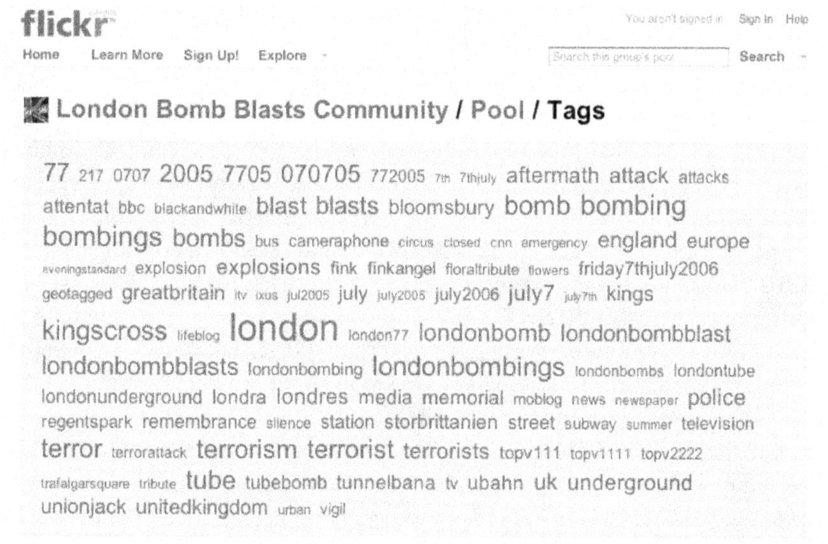

flickr

You aren't signed in   Sign In   Help

Home   Learn More   Sign Up!   Explore ·   Search this group's pool   Search ·

London Bomb Blasts Community / Pool / **Tags**

77 217 0707 2005 7705 070705 772005 7th 7thjuly aftermath attack attacks attentat bbc blackandwhite blast blasts bloomsbury bomb bombing bombings bombs bus cameraphone circus closed cnn emergency england europe eveningstandard explosion explosions fink finkangel floraltribute flowers friday7thjuly2006 geotagged greatbritain itv ixus jul2005 july july2005 july2006 july7 july7th kings kingscross lifeblog london london77 londonbomb londonbombblast londonbombblasts londonbombing londonbombings londonbombs londontube londonunderground londra londres media memorial moblog news newspaper police regentspark remembrance silence station storbrittanien street subway summer television terror terrorattack terrorism terrorist terrorists topv111 topv1111 topv2222 trafalgarsquare tribute tube tubebomb tunnelbana tv ubahn uk underground unionjack unitedkingdom urban vigil

FIGURE 4.3 Tag cloud for the London Bomb Blasts Community (Flickr 2008). Image by Yahoo! Inc. ©2012 Yahoo! Inc. Flickr and the Flickr logo are registered trademarks of Yahoo! Inc.

so, they define hashtags, create Facebook groups, build Flickr photo pools, add content to Wikinews pages, and comment on mainstream media news reports.

Much of this work on Flickr was done through tags, but it was also done through comments on the images themselves. While we can see problems of language stabilization discussed earlier, another problem was fracturing of meaning across the single pool. Were people talking about bombings or blasts? For New Yorkers, a "blast" is a fun party. For Londoners, a "blast" is an explosion. Even naming the event proved difficult: is it the "londonbomb" or the "londonbombblast"? These kinds of language slippages can cause problems for automated systems, as I discuss later. Suffice it to say, the need for human anchors remains an important factor for understanding that an event is occurring, that it is valid, and that networks need to start forming around these human anchors to help push information through the network.

Looking specifically at the image of Adam Stacey, we can see how anchors and participants began to define the event of the London bombings and to become useful to each other. This image of Stacey escaping one of the bombed areas appeared across multiple websites, eventually leading to its inclusion in mainstream media news reports. Looking specifically at this content, we can focus on two specific instances of its use to understand how this data transformed into information. The grainy photograph appeared on a mobile blog (a moblog), and a different participant later uploaded it to Flickr. In both spaces, participants were able to comment on the validity of the image, adding more content as they learned more about the event and the image.

Before the London bombings, Alfie Dennen maintained a moblog that could be accessed and updated via mobile device as well as desktop computer. Though we commonly use smart phones in this way now, the ability to connect a mobile device to a blog and update the blog while on the go was still fairly new in 2005. Smart phones now make this process a standard practice among many social web participants, especially on sites like Twitter and Facebook. For Alfie Dennen, such a connection meant that he could update his personal blog, *Alfies Moblog*, just minutes after the attacks. Figure 4.2 shows an image taken of Adam Stacey that Dennen posted to his site. As Dennen (2005) describes it, the image shows "people trapped in the tube." So began the moment of problematization as Dennen did his best to become useful to the community by sharing this image with his moblog followers. The photo of Stacey circulated throughout the internet, eventually making its way to television news reports around the world and into the pages of newspapers worldwide in addition to being reposted by many others across the social web.

The actors, or grouping of actors, in this assemblage include a variety of people, technologies, and organizations: Adam Stacey, Alfie Dennen, *Alfies Moblog*, cell phones, computers, Flickr and other social web tools, and *BBC News*. Additionally, many other participants took part in distributing this image, adding information to its narrative and adding it as information to other narratives. On *Alfies Moblog*, the initial postings that other participants made in response to the image were from a group of people with already-established connections to one another. Yet, as these participants began to take the image beyond Dennen's moblog through linking to and reposting the photograph, their efforts recruited new actors into the assemblage. Other people, websites, and organizations constructed new connections for the data, adding further context and stabilizing the narrative as a result.

A participant who was not the owner of the image in figure 4.2 uploaded it to Flickr (see figure 4.4). Four important pieces of content appear here. The first is the poster's referencing of the image's original post on *Alfies Moblog*. He refers to the original in the description of the image on Flickr, adding a link to the source. In doing so, he translates this piece of data into information, confirming the source material and thus the event itself. Also important in this transaction is the voice of an anchor, Flickr participant fgt, who leaves a comment requesting that the image also be placed in a photo-pool community. That anchor, David Storey, is a key actor in this network, as he works to establish the initial network and then to stabilize the network, as we see in the next section.

We can also tell that this image was taken on the day of the bombing, 7 July 2005 (Applegate 2005). Of course, this date is when it was taken from a website, not taken in the sense of an image being captured on a specific day by a camera, which is what "taken" ordinarily means in the Flickr user interface. This date is

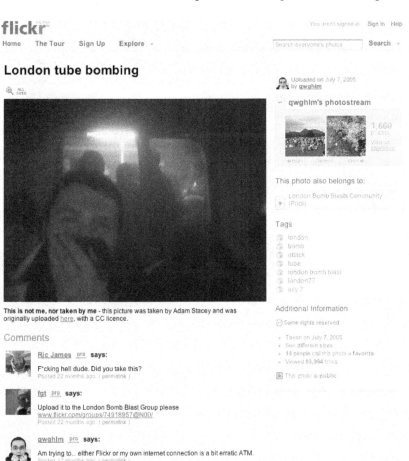

FIGURE 4.4 Image of Adam Stacey on Flickr (Applegate 2005). Image posted to Flickr by Applegate under a Creative Commons 2.0 license at http://www.flickr.com/photos/qwghlm/24230239/. Original image published under a Creative Commons 2.5 license by Adam Stacey.

important, as its proximity to the time the event occurred also helps validate the image as part of this event. While none of these pieces of evidence is solid in and of itself, when the pieces are taken together, participants can validate this image as real information about the event. Other key content in figure 4.4 are the tags that the original poster used to describe this image: "london," "bomb," "attack," "tube," "london bomb blast," "london77," "july 7." These pieces of data help with problematization because they help define what is occurring in this event—a

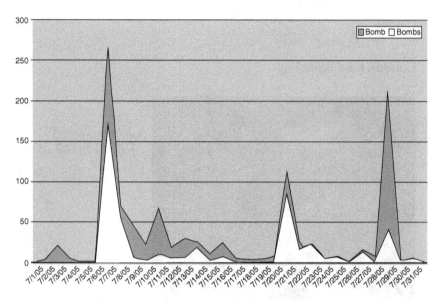

moment early on in the event before language stabilizes. Using this many tags helped other Flickr participants locate the image by using Flickr's search function. Clearly, these tags were important for participants to communicate with each other. In figure 4.5, we can see that a spike occurred in the use of the tags "bomb" and "bombs" on the day of the London bombings, as well as later in the month when attempted bombings occurred in London on 21 July 2005.

During this event, an actor network coalesced around the image of Stacey and then dispersed across multiple people, groups, and technologies. The diagram in figure 4.6 reveals the major actors (such as terrorists, Alfie Dennen, blog participants, camera phones, Adam Stacey) that assembled around this fire space, as well as the type of relationship each actor shared with the mutable mobile at the diagram's center, the photograph of Adam Stacey. These other network actors added key pieces of information concerning Adam Stacey, other possible victims, and the London Underground, all in an effort to provide more context for the scene shown in the image. While participants were discussing the photo on *Alfies Moblog* and Flickr, other participants sought information concerning the photograph to take to other spaces such as Wikinews. The lines in figure 4.6 refer to the need for certain actors to be present in order for the central artifact to exist as the mutable mobile. From the perspective of the architect and the researcher, the purpose of the image is to distribute information and make people aware of what is occurring. The moblog and blog participants are necessary to generate

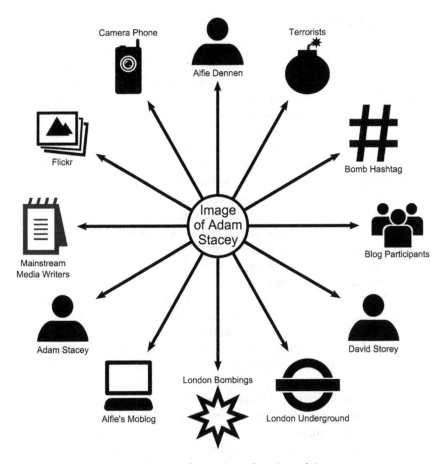

FIGURE 4.6 Actor-network diagram for tracing Adam Stacey's image.

interest in the image and eventually push the content outward to other parts of the social web. For the wider audience, the mainstream media are necessary to spread this content.

In this example, I use ANT to situate specific interactions within a larger set of connections, which can contextualize, recontextualize, and situate the content that individual articulations may form or uncover. While these actors cooperate to move this content throughout the network, they become useful to each other, negotiating meaning and providing support as content contributors, anchors, and moderators. They can do so if the system permits this kind of activity, which at the time of the London bombings, Flickr supported. The larger goal of these moves is to find enough data to provide context and stabilize a narrative of the event so that participants could validate the data as information.

Stabilizing the Network

Across the four locations where the bombings occurred, people communicated with others through cell phone data streams and connected with others online. Social websites such as Flickr, Wikinews, Internet Relay Chat (IRC), and the "Have Your Say" section of *BBC News* articles became locations of information coordination as everyday people pieced together information about this catastrophe. Over and over again, the definitions of the disaster—what was happening, where it was happening, and who it was affecting—were repeated. During this stage of translation—interessement—the anchor actors brought clarity to the event and encouraged other actors to agree to these definitions. Among these spaces, social web participants used Flickr extensively. Thus, it was one of the first spaces in which actors stabilized. Though originally created as a space where photographers could share their most recent photos, Flickr instead became a place to upload and catalog images of disaster, a place where actors invited others to help create narratives of a major city's terrorist attack through comments and by generating pools of related images.

On Flickr, participants can organize individual photos and photo pools however they choose. They can also create personal pools where they can organize just their content. Pool owners can arrange images to chronicle their experiences (Cridland 2005). In practice, owners can rearrange these images in any order, whether chronologically, randomly, or according to some other scheme deemed most suitable. Personal photo pools like these are common, and numerous personal pools documenting the London bombings quickly appeared across Flickr. Flickr participants can also share the images they have captured with public photo pools that communities maintain. To do so, in many public photo pools, such participants must request a membership with the community. Conversely, one of the photo pool's owners, or even a volunteer, can invite specific individuals to add an image to the public pool. As a result, images posted to Flickr, including those capturing the events of 7 July 2005, belong to multiple photo pools, both public and personal. Once invited to or approved for membership in a public pool, participants can add or remove images from the photo pool at any time. Although this system is an excellent example of how folksonomies can help participants organize content, an official structure, a taxonomy, would allow participants to easily locate images as discussed earlier, creating a common language that could help unite content. Thus, pooling images into large public collections was a vital means of moving information among Flickr participants after the London bombings.

As I discussed earlier, the London Bomb Blasts Community created the largest of these public photo pools. The activities of this community, especially those of the anchor of the photo pool, demonstrate participants' use of different tools. Once actors verify this information, they can then wrap context around the information to make it useful to others as they redistribute it as community knowledge across these networks. The London Bomb Blasts Community (2008) had as many as 427 members at one time and has been the repository of more than 1,100 photos related to the 7 July 2005 attacks. Although the user interface has since been updated, accessing the photo pool's home page in 2005 would automatically cause the six most recently uploaded photos to display. By clicking on the pool link, visitors could view the photos in a chronologically ordered gallery as people uploaded them. Visitors could also navigate multiple gallery pages using the numbered links at the bottom of the page. These features meant that participants posted images on Flickr that not only provided information about their safety and well-being but also spread information concerning related issues such as transportation difficulties and events taking place along their routes home.

Many images posted to Flickr included emergency personnel and police officers directing people away from the bomb sites (Bradshaw 2005; Francis 2005), emergency vehicles racing to and from the scenes of the attacks (Cronin 2005; Stott 2005a; Tinworth 2005), helicopters flying over the city (Moggan 2005), and armed officers in boats watching the scenes from the River Thames (Stott 2005b). All of this definition making is what marks interessement, the stage of translation at which anchor actors work to stabilize the network. The widespread use of mobile devices such as cell phones among commuters meant that people captured photographs of the immediate chaotic aftermath. Such images showed people fleeing train carriages and individuals evacuating the London Underground system (Applegate 2005). Similar to those watching television and capturing images of the broadcasts to upload to internet sites (rockmother 2005), participants were also creating screen captures of the *BBC News* website to upload to social websites (Roberto 2005).

Through this mix of activities, Flickr became a hub for these assembling actors. Both people and technology could connect the images and comments within Flickr itself to information contained elsewhere, although they could only participate to the extent that Flickr and the users there permitted. Not all content owners permitted the tagging of photos and leaving of notes on images, but many more allowed comments and links to be made among content pieces. Networks expanded and contracted to both take in information and spread information outward from Flickr. Thus, Flickr became part of a larger ecosystem of activity across blogs (London Bloggers 2006), IRC (Ito 2005), and news sites (*BBC News* 2012).

Pools like those that the London Bomb Blasts Community created actually generated the organic narrative of the London bombings on Flickr. Public photo pools, like the one that the London Bomb Blasts Community created, often focus on a topic, event, or common theme. Much like with the creation of groups or public fan pages in social networking sites such as LinkedIn or Facebook, Flickr automatically designates the person who originally creates a public Flickr pool as the owner of the pool. The pool's originator can then designate others as moderators who help manage the image pool as it continues to evolve and expand. The two main tasks of these anchor actors are to find images relevant to their photo pool and invite the owners of these images to add them to the public collection and to continuously monitor the pool to ensure all images are relevant to the collection—in the case of the London Bomb Blasts Community, the pool's needs.

After the attacks, these images focused on the exchange of crucial information that the community deemed necessary. Later, the pool became a repository of photos that captured tributes to or memorials for the victims. Understanding how anchors find, aggregate, and sort images as well as managing the participants of other actors who are themselves seeking out relevant information are crucial steps for researchers and architects of social web tools. Figure 4.7 shows an ANT diagram with the London Bomb Blasts Community as the center artifact. Each connecting actor must be present for this central artifact to form and then remain relevant to the network.

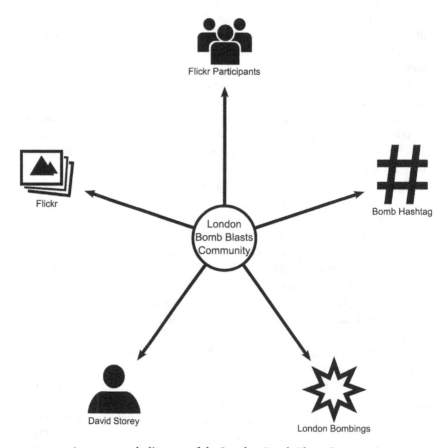

FIGURE 4.7 Actor-network diagram of the London Bomb Blasts Community.

The anchor, creator, and originator of the London Bomb Blasts Community, David Storey, would locate images on Flickr that he thought were relevant to the group's purpose. As numerous Flickr participants posted photos to their own photo streams, Storey would leave comments on relevant photos asking if the photo's owner would add it to the London Bomb Blasts Community pool (Storey 2005b). Storey would also leave a link to the group's main page to facilitate finding the group and adding an image to the pool. Storey's actions were specific moments of articulation in which he linked his group pool to individual participants and created connections that could possibly help the larger community of participants. This connecting of images, pools, and participants also provided avenues for further articulations by other participants both within the London Bomb Blasts Community and outside the membership of the group, as well as with other spaces entirely outside Flickr, such as blogs and wikis. The work of anchors and other Flickr participants to create these assemblages reveals the need for flexible, adaptive systems that facilitate the kinds of linkages and movement of information so important to networks responding to disasters.

David Storey and the London Bomb Blasts Community could leverage a participatory space with an ecosystem of participants comprised of regular, everyday people that the London bombings directly affected. Storey's efforts were essential to collecting and aggregating images for the London Bomb Blasts Community. His work enabled other participants to more easily locate information within the pool of photos and to add other pieces of information to photos in the pool. Through his efforts, Storey made himself a content aggregator, performing the work of a search engine or RSS feed as he located relevant photos and encouraged others to add their images to the community. Storey's case points to the necessity of participants willing to take on the task of searching for, gathering, and managing actors within a network, whether these actors are images in a collection, interactions with technologies, or participants within a community or group. Anchors such as Storey are examples of the effectiveness of people-powered tools and content as communication tools. However, tools that could have intelligently performed some of this work would have significantly helped him.

The image of Adam Stacey provided information that was useful to participants more directly concerned with questions other than Stacey's status, such as who was caught in the middle of London's 2005 terror attacks and what was their condition. The actor network that emerged around the photograph of Stacey also supported a number of other networks that used the image as data to contextualize other narratives. This example is a crucial maneuver that experience architects need to consider as they build systems for social web participants to move information. They must not only consider the technologies involved, the contexts of use, and the ways participants grab content and bring it back to the space that immediately concerns them; rather, they must also be aware that other participants and other networks are operating simultaneously. And while an image such as that of Stacey is a focal point for many actors, its function as a mutable mobile also means such an image can provide much-needed data triangulation to other networks.

Practical Solutions

So far, we have examined how participants defined these events, how volunteers moderated content, and how the network stabilized. We have seen through these two moments of translation (problematization and interessement) how content can move from data to information. In addition to creating better support structures for human participants, we need to think about how we can aid in validating information through various technological applications. This section discusses some of these solutions in regard to information validation.

Diagrams such as the ones in figures 4.1 and 4.6 are useful because they can help identify actors and the types of connections to other technologies that the actors support. In addition, these diagrams illustrate the articulations that actors can maintain within each space. A blog allows visitors to comment and leave links or messages if they choose to do so. If a participant finds something relevant to others frequenting that blog, that participant can leave behind the link and a brief description of what others may find at the link's destination. If information is inaccurate or outdated, then participants can add comments to warn other participants. On the other hand, a community can collectively maintain a wiki so that outdated or untrustworthy information can be more thoroughly managed or even deleted. Some systems support tagging that makes content more readily searchable and sortable, while others do not. Though the technologies themselves do not necessarily support moving content in fluid ways, human actors still allow this movement to occur as quickly and effectively as possible. As a result, participants can maintain the stability of information but may sometimes experience difficulties in doing so. Understanding the connections formed within the network aids experience architects as they work to develop systems that better support communication within these fire spaces.

Key to these actor networks are anchors like Storey. In the next chapter, I discuss another anchor who was also key to a network. Working with these anchors is important for system owners and content producers. These anchors attempt to collect content, make it available for others, and help track down important information. Recognizing these anchors for their work is important. Using badges on these systems can allow others to recognize their work. Giving anchors special access to content can help them get their jobs done, and special dashboards can help them monitor content during these events. We also need to build software and solutions that can facilitate anchors' searches for photos, links, and relevant data.

Cataloging the thousands of tweets, images, maps, news articles, and other materials generated after an event is often impossible. Missing information and misinformation have the potential to hamper a community's ability to validate content. Technologies can potentially perform simple tasks that occupy dozens, hundreds, or even thousands of hours of time for people, but anchors who understand the context, content, and concept of what other human actors are trying to accomplish in these networks need to manage these technologies. Such systems could also free up anchors to pursue other more difficult tasks for the community. An example of such a system task is searching for relevant photos and intelligently aggregating them.

Only by working within these communities as either participants or anchors can architects fully understand the needs of the community as a whole and those of individual participants who take on specific roles to help the group in its collective efforts. Experience architects and researchers can make such connections during these events. We can improve these systems by watching how anchors work, as well as how participants validate data and present information. But we can only do so by acting as participants ourselves and involving participants in building improvements and developing innovative solutions.

Another issue that can affect anchors during disasters is speed. In cases such as the London bombings, the process of sharing speeds up. This event not only proved the utility of the social web for finding, collecting, and distributing information but also brought to the forefront how participants on scene could use mobile technologies to collect information and post it to the web almost immediately after an event. Systems can help stabilize language by connecting different actors and then anchoring the collections, allowing content to transform into knowledge. Humans can help validate these findings, deciding when "blasts" means parties and when it means terrorist attacks.

Another key issue is validating the location of these blasts. One example of how Flickr could be used to help pinpoint location is seen in figure 4.8. This image is a photograph that Flickr participant John Howard (2005) took with a camera phone. It is a seemingly random photograph taken during the morning hours after the London bombings. Howard states that he took the image "before they [the police] started moving us back." The emergency personnel, the landmark of St. Pancras adjacent to King's Cross station (in the upper right corner), and the scene of evacuated commuters that the image contains validate the authenticity of this image. This photo is also noteworthy as a view into the sense of community created by the commenters who thanked Howard for posting this image. Locating this image in the pool of photos, we can see how the image is part of the London bombings event because of these distinguishing markers.

In some cases, participants on Flickr allow their location metadata to be uploaded when they add images to Flickr, and such metadata can be altered and thus become misinformation. Using a tool such as Google Earth, we can compare what we see in the photograph a participant uploads with that area of the city, further validating this content. So how can we automate some of this validation and provide tools for other actors to perform this level of validation? One way

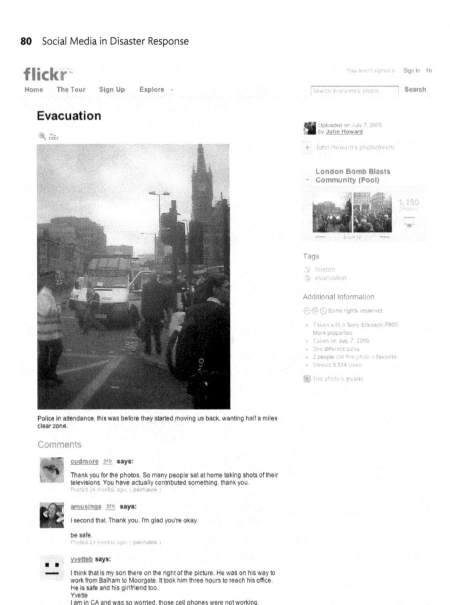

FIGURE 4.8 Flickr photo of commuters in London on the day of the bombings (Howard 2005). "Evacuation" by John Howard, reprinted with permission and available at http://www.flickr.com/photos/jmh/24223193/.

would be to link GPS content in ways in which everyday people can confirm location, perhaps by linking to Google Earth and using Keyhole Markup Language mapping solutions. Any automating solutions must also allow for human intervention as a form of fail-safe for these technological fixes.

Looking at the image in figure 4.8 from the perspective of triangulating technical information, we can learn a lot from image quality and metadata. The photo is rather grainy, suggesting that it might have been taken with a camera phone. When uploading from a camera—phone or otherwise—Flickr captures metadata from the photo's file. This information (e.g., "Taken with a Sony Ericsson P900") is then listed under the additional information section in the user interface (Howard 2005). By displaying this information, Flickr also provides tangible evidence of the kind of camera that took the photograph. Such metadata can validate information. Information about the type of tool that the participant employed validates the authenticity of the image as well as the phone's owner. It can also tell experience architects and researchers more, such as what cameras are popular in certain areas, allowing them to build systems that can, for example, connect with camera features. Both actors—the camera phone and the person using the phone—are necessary components for creating and strengthening this assemblage; both need to be validated for authenticity.

Conclusion

Understanding how networks take shape and stabilize is important for researchers and experience architects. We cannot research these phenomena nor build for these experiences without knowing how translation occurs. By tracing the process through problematization and interessement, we can locate how these networks stabilize and how actors become useful to each other. Through tracing, mapping, and describing these examples, researchers can better see what participants require to transform data into information (Potts 2009b). In doing so, we can improve these sociotechnical systems. As researchers and practitioners, we must architect for these experiences rather than over and around them. By looking across these networks to document relevant technologies, organizations, and people, architects can better plan further implementation of technology.

For everyday users, navigating the social web can be confusing; a disaster magnifies problems. In making these communities visible, architects can more readily see issues related to participatory experience. These cases and diagrams can help show stakeholders that they should support effective and efficient communication systems based on use rather than constraints, systems that allow practitioners to solve problems that affect usability during everyday experiences as well as during times of crisis.

Moreover, we can leverage the voluntary work of social web participants to more efficiently accomplish tasks that many organizations might traditionally perform with heavily strained or inadequate human and financial resources. Yet, for the moment, human actors simply make do with the resources they have—resources that are not architected for this type of work and that often hinder participants as they perform their tasks. In the next chapter, I present the Mumbai attacks as a transformative moment. Through the efforts of a small group of participants serving as anchor actors, participants across the social web assembled to collect, share, and verify information about victims of the attacks. Using systems such as Google Docs, blogs, Twitter, and many other social web tools, they developed a robust actor network that serves as a case for future research and architecture work.

Using ANT can help researchers imagine ways in which methods and theories can transcend their professional community and be useful for those in other knowledge-making areas, including the social web. By making such moves, experience architects can continue in a long-established tradition of advocacy. Although these research areas reside on the borders of many disciplines, these methods can remind practitioners and researchers that their work—and the way they make knowledge—brings new perspectives on existing problems. As Stolley (2009) noted, these "technical and theoretical foundations for realizing such activity are already in place" and, therefore, we "need only to make purposeful, activity-driven use of them" (370). In the next chapter, we look at one example of this kind of activity where participants use social web tools in innovative ways.

5

TRANSFERRING KNOWLEDGE DURING THE MUMBAI ATTACKS

In late November 2008, a series of terrorist attacks occurred across the largest city in India. Known as the Mumbai attacks, the attacks targeted 12 locations, including hotels, a railway station, a hospital, a religious center, and a café. The terrorists attacked multiple sites in a coordinated effort and left 166 civilians dead and at least 304 injured (Chief Investigating Officer, Government of India 2009). Because the terrorists dispersed the attacks across the city, Indian authorities faced incredible difficulties finding and verifying information about victims. While this event was erupting, the Western mainstream media were also struggling to keep up. Over the duration of the attacks, from 26 to 29 November 2008, social web volunteers worked to locate data from various sources, collect and validate the data, and then redistribute information about the hundreds of people left dead, missing, or injured.

Reaching this level of knowledge production signaled a watershed moment for social web participation. Although reporters had yet to cover the attacks, volunteers were writing about this event, sharing details about the attacks, and distributing content across multiple systems. In what was otherwise a vacuum of information, the social web became a critical tool for aggregating, organizing, and disseminating information during the event and its aftermath. The Mumbai attacks of 2008 provide another example of information coordination among social web actors, who pulled together to "act as *one piece*" (Latour [1987] 2003, 129; italics in the original). Desperate for information, participants collected and organized information about the injured and the deceased, helping families and friends locate their loved ones. For example, using tools such as Twitter, blogs, and Google Docs, participants documented the injured and dead, bringing answers to many. Moving across multiple systems, volunteers across the globe distributed information and shared knowledge.

This chapter describes methods for tracing participants as they form actor networks, validate information,[1] and push knowledge through the network. So far, we have focused on the first two of these three stages. Here, we shift our focus to the final stage, showing how people and technologies transfer knowledge by helping to spread information across the network. We look specifically at how volunteers effectively share knowledge across the network, liberating this information from the walled gardens of closed systems. Focusing primarily on occasions when actor networks stabilize, we examine how participants take part in these ecologies and mobilize the network. In this mobilization, they can fully realize the actor network, pushing knowledge outward to people and technologies during times of disaster. Understanding how networks operate during these moments shows us ways that technological and policy solutions can support the work of these participants.

Building upon our earlier discussion of how content moves through the three stages of data, information, and knowledge, I show how these activities take place on the social web. By observing and participating within social web spaces, researchers can begin to understand how these systems work, and experience architects can learn how to develop for them. In these spaces, volunteers want to participate. Sometimes, these spaces support participation, but more often, systems and policies stunt opportunities to participate. The examples in this case demonstrate what happens when systems involve a space for participation. Although improvements are still needed for these technologies and processes, we begin to see what a future might look like if we create more participant-centered architectures. Looking at the final steps in how networks mobilize and gain participants can help researchers and practitioners create better systems and policies to support such participatory spaces.

Ecosystems and Knowledge

Actors are people, technologies, groups, and organizations that form networks to push information through sociotechnical systems. Across the social web, actors locate content, confirm details, and supply information to their communities as validated knowledge. While Latour (1999b) described these actors as having equal agency as part of actor-network theory (ANT), the examples in this case study point to human actors and the groups they form as having far more agency than the technologies deployed around them. That said, we must understand that these technologies support human work, that they can aid in organizing content, and that they can aid in pushing content across the network. So, while humans have more agency than the technologies they deploy, these technologies can be helpful—if not essential—in propelling knowledge across the social web.

Phases of Content Transformation

Data are simply content. For example, data can be tweets on Twitter or images on Instagram. Data are content without context and meaning. Data become information when an actor confirms the data as valid. In chapter 4, I explored several examples of validation, such as confirming that an image of a disaster is from the city where the disaster occurred. In turn, information becomes knowledge when it is redistributed to the community.

Content is transformative. An example of how content moves from data to information to knowledge can be seen as actors work across systems. A person may see a tweet on Twitter that states a major brand has been sold. To confirm this news, the participant might search for the news on Google, *BBC News,* or another mainstream news site. The participant may also go directly to the source, looking for clues on the brand's website. Once the actor validates the information, the actor can then retweet the news and link to the source material to validate the original tweet. By redistributing the original content with the validated source material, the participant transforms the information into knowledge and closes the circuit. The network on Twitter now has knowledge of the event, validated information, and the data to support it.

As this example shows, the distribution of knowledge does not occur when actors are idle. Rather, knowledge emerges from the mobility of information. Participants need to be able to post information, reply to reports, leave images, and tag content. In some cases, the actors are human anchors passionate about a subject area or desperate to help locate content. In other cases, the actors are technologies leveraged to help move information through groups. These anchors help form networks, stabilize them, and carve out pathways for communication among participants. Recent research points to the need for mobility across these social experiences, citing Twitter in particular as a space where this mobility is taking place. The ability to move information across a network is crucial for social web participants (Potts and Jones 2011). Without such mobility, we are trapped in walled gardens, unable to share knowledge with the wider network.

Chapter 4 outlined the first two moments of translation (problematization and interessement), during which the anchor actor appears, the network stabilizes, and actors are encouraged to accept the anchor's definitions within the network. These definitions can be something as basic as agreeing that an event is occurring, or they can be something more complex, such as determining which hashtags are relevant to the event. During these early stages, anchor actors appear, ready to volunteer and organize content. For example, participants on Flickr can tag images using the word "Mumbai" so others can then locate content about the city. Such definition creation via group tagging stabilizes a network. Actors can confirm the content in the images by leaving comments on these images or by linking to relevant news articles or associated content on the web, such as lists of missing persons.

Unfortunately, the systems in which such interactions take place too often stall mobility by restricting use, building for content containment, and disallowing participation. After Hurricane Katrina, CNN's website was a walled garden, forbidding everyday people from altering content. Only CNN agents could modify its missing-persons list; people could do nothing more than email concerns to the website producers. Future systems must take into account the freedom of exchanges in support of a participatory culture. By analyzing how content moves through stages, researchers and practitioners can better understand how these activities are taking place. Such an understanding can lead to improved experiences, policies, and systems.

Actors Working Better Together

Using the framework provided by ANT, researchers and practitioners can pinpoint the interactions among people, technologies, systems, and organizations. Translation as defined by Callon (1986) is a key factor for analyzing how actors pursue their work: translation is the process by which an actor network is punctualized—formed, stabilized, and activated. Punctualization is a simplification; for example, we talk about a television without having to mention the cables, plugs, and tubes that come together to create that television (Law 1992, 384–85). Although "punctualization is always precarious . . . and may degenerate into a failing network," it offers "a way of drawing quickly on the networks of the social without having to deal with endless complexity" (385). Employing ANT to examine experiences is useful because it allows the experience architect to map participants across many technologies. Mapping the actors available to these participants allows architects to create systems that encourage information movement across these systems, supporting actor use and user interfaces that enable these activities.

This chapter outlines the final moments of translation, showing each process in relation to specific examples of people turning to the social web to volunteer and to find information and help. These final stages are referred to as "enrollment" and "mobilization." During enrollment, actors willingly join with other actors, aligning themselves with various anchor actors and collections. During mobilization, actants (or groups of actors) form a network that can accomplish more than separate, individual actors could and can thus reach punctualization. Through translation, the actor network gains coherence and purpose, strengthening its ability to pass along and validate information and then, finally, to redistribute this knowledge across the network.

Here, I discuss each of the earlier moments of translation to provide background for a deeper examination of the later stages. For problematization, we examined how actors determine what problems exist within these events and how actors work to become anchor actors or participants. We then examined interessement, how anchors help stabilize participants and encourage the gathering of information. Enrollment, the act of validating content, occurs when anchors track down details regarding information participants. By visualizing networks and seeing the actors involved, researchers not only learn who is participating but also where they are participating, what tools and systems they are using, and what other technologies are further supporting their efforts. With this knowledge in hand, researchers and experience architects can update existing systems by adding features that better support the flow of information, especially during moments of crisis or disaster. Not only do researchers need to explore and understand the roles of anchor actors working to find and gather information, but they should also support the instances of composition, collaboration, and coordination as actors collectively work to respond to their own needs or the needs of others. This collective response is the key outcome of translation and is enacted in the final moment: mobilization.

Enrollment: Anchoring to Collectors, Concentrating Activity

The third moment of translation is enrollment, when actors are "willing to anchor themselves to the collectors" of data (Callon 1986, 211). We recognize this moment by the actors' acceptance of the definition of their network. This accepted definition, created during the problematization stage and made cogent during interessement, further intensifies the obligatory passage point as actors strengthen the central network by accepting the anchor actor's definition of this space. Enrollment is made visible through various movements. By adding information to a community based on an anchor's request, actors attach themselves to the network and to the anchor actor.

As work increases, participants who visit these spaces see an increase in the information in the stream that belongs to the community, and this activity can lead to a form of mobilization. As participants notice photos added to a social stream, enrollment encourages them not only to view the stream but also to add their images to the community. This activity was evident in the wake of the London bombings, as discussed in chapter 4. Such concentration of activity can encourage further participation, allowing us to observe how participants work together to communicate about events. Architects must uncover these mediated interactions among anchors and actors in order to know which interactions to facilitate through new social web systems.

Mobilization: Networked Action

The fourth moment of translation is mobilization. During mobilization, the actors marshal their peers to action. Callon (1986) described this moment: "Through the designation of the successive spokesmen and the settlement of a series of equivalencies, all these actors are first displaced and then reassembled at a certain place at a particular time" (217). Here, Callon describes how actors are assembled; these assemblages are constructed to push information through these networks.

Many diverse cases reveal how these assemblages occur online. By examining the use of the social web during times of disaster, researchers can benefit from the high-pressure, intense usage patterns that such examples show. In the aftermath of the Mumbai attacks, interconnected actors created assemblages through a multitude of social web tools such as Twitter, Google Docs, and blogs. Amid the chaos of exchanging information and discussing the disaster, participation across the social web spiked.

In the world of software and the internet, concepts of place and time are virtual and somewhat disjointed. Geographic location can be either a hindrance or help. For example, time zones are not always a clear indication of volunteer availability. Within the context of online social networks, time zone changes are less about losing touch with other actors than they are about ensuring that work can continue.[2] While one location sleeps, the other location works to validate content and distribute knowledge. Of course, in the case of these cross-cultural issues, distribution of work can lead to inaccuracies. But the involvement of many participants and anchors can lead to a faster pinpointing of data and a quicker distribution of knowledge.

During a live event, the urgency for information is supercharged, whether the urgency is to locate missing persons, track down stolen cars, or discuss a keynote speaker. While the stakes may differ, the needs of participants remain paramount. Participants work to quickly locate data, validate that data as information, and push the information through the network as new knowledge. This participant urgency can lead to the construction of innovative solutions as participants work to build coherent narratives about these events and send them into these spaces. The technologies participants leverage must handle these exchanges, or else participants will repurpose digital spaces in ways the system architects never intended. That latter part is not a problem provided that these spaces are underdesigned in ways that allow for this kind of participation. By locating these activities, we can begin to understand the needs of the participants. This understanding can help us figure out how to construct systems that can support flexibility, encouraging new kinds of use.

Distributing Knowledge across Systems

Throughout several harrowing days, the Mumbai attacks enrolled social web participants across the world. Participants raced to help locate the missing and validate information about the attacks. The media were often dependent on participants to relay information using Twitter, YouTube, blogs, and other social web tools. Media coverage focused on the events at the Nariman House and the Taj Mahal Palace hotel. To the people of Mumbai, the hotel is a place of great cultural heritage ("CNN-Dina Mehta.avi" 2008), and it is also a well-known icon for people outside India. The Nariman House is a Jewish community center, or Chabad house. Participants extensively leveraged Twitter to communicate about the disaster. These participants' work advanced the use of the social web as a space for communication during times of disaster. The Mumbai attacks saw the use of numerous social web technologies. Nimble and enterprising social web participants quickly outpaced both the press and official government organizations in disseminating information. Participants quickly turned to social web systems for gathering information about the attacks. People used technologies such as Twitter to communicate what was happening and to provide information about their well-being. Shared writing spaces such as Google Docs became a valuable tool for participants needing to collaborate, coordinate, and contextualize information. And the participants completed all of these activities at a pace much faster than traditional broadcast media channels.

After a disaster, the sense of urgency is high and the need for information to flow through new networks is critical. This section discusses examples that illustrate how anchor actors enrolled participants and pushed knowledge through the network to share it with the community. These examples illustrate some key points about how actors perform these acts and distribute validated knowledge. We can apply such examples to everyday communication issues, encouraging the development of social web systems that provide flexible and contextualized experience.

As with prior events, after the Mumbai attacks, the websites on which people focused their energies were those for which they had established networks— spaces where they already spent their day-to-day life before the event. Again, anchor actors were unprepared at first and then quickly attempted to react to participants who were stretching these systems in ways the designers had not imagined. The best hope for success lies in these reactions. As obligatory passage points for information, the anchor actors were necessary components in the network. By providing ways for anchors to engage further with these communities, architects of social web tools can have a greater chance for community success— even when that community is as temporary as it is during times of disaster.

One of the key strategies anchor actors deployed during the Mumbai attacks was the coordinated use of different social networking systems and collaborative tools, including Twitter, Google Docs, fax messages, and mobile phones. Participants communicated, recruited other participants, and tracked down information using Twitter. Some participants shared information about possible carjackings (scorpfromhell 2008); others shared emergency numbers (pradx 2008b) and even retweeted personal numbers to help coordinate information (krisnair 2008). Figure 5.1 illustrates one of these examples, as a Twitter participant uses the agreed-upon hashtag for the event and provides a link to the *Mumbai Help* blog. Participants used multiple hashtags to connect actors (SSSIndore 2008). Thus, these hashtags became intermediaries that could help direct conversations across the flood of discourse occurring within Twitter.

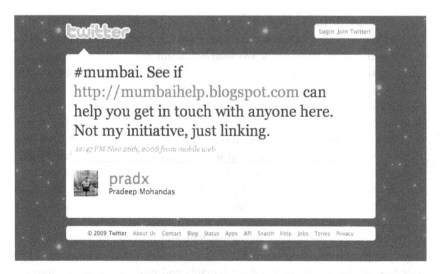

FIGURE 5.1 Example of coordination on Twitter during the Mumbai attacks (pradx 2008a). Tweet by pradx; image printed with permission and available at http://twitter. com/#!/pradx/statuses/1025211644.

Participants collaboratively organized and edited information that had been collected into a public Google Docs spreadsheet (Mehta 2008b). They then turned to the social web to trace bits of evidence, cross-referencing data, verifying what was valid and identifying what was still questionable. In this case, participants used the Google Docs spreadsheet to inscribe shared knowledge across the actor network to reach punctualization.

In the next section, we dive deeply into an example of how participants realized punctualization. As networks concerning the Mumbai attacks coalesced, actors assembled into a group that could accomplish far more than the sum of the actors' individual tools and abilities suggested. The anchors drew together multiple social websites and coordinated participant efforts to perform vital knowledge work that would have otherwise taken days for official channels to complete. Their efforts connected people, information, and stories to provide much needed assistance that aided victims, the families and loved ones of victims, and emergency response personnel.

Enrolling Actors and Exchanging Knowledge across Ecosystems

Like other crises, the Mumbai attacks presented unique challenges to participants leveraging social web tools to perform valuable knowledge work. In this case, the simultaneous attacks led participants to coordinate information across a number of online systems and information emerging from a myriad of different sites around Mumbai. Terrorists struck several major locations within the city, spreading the efforts of emergency responders and mainstream media. Whereas our previous examples from disasters in London and New Orleans highlighted separation between the work of participants in the social web and that of mainstream organizations such as news media and relief agencies, the events in Mumbai displayed a high level of coordination among members of the media and social web participants. By relying on numerous social web tools, these actors coordinated their efforts to collect, organize, and verify information in rapid order.

By stepping into the fray, social web participants led by an anchor actor filled the information void quickly and effectively. Locating the missing during a disaster and in its wake is one of the most critical activities that human actors face. Participants reached the first two stages of translation (problematization and interessement) rapidly, although somewhat chaotically, during this disaster. Moderators worked to identify the event, exchange information, and stabilize hashtags. After the actors agreed that a disaster was occurring and came together to stabilize their network, the latter stages of translation—enrollment and mobilization—began. Then, hashtags were used to enroll actors and encourage the exchange of key information about the bombings, emergency contacts, and a possible hijacking. Twitter was a central location for activity, with images and video coming out of Mumbai from everyday participants before news organizations had organized on the ground. Supporting the network during this event involved many anchor actors, some of whom had prior experience coordinating information together during prior disasters ("CNN-Dina Mehta.avi" 2008; Mehta 2008a). Work was decentralized but organized, allowing actors to "mobilize in self-organizing kind of ways, in bottom-up ways" ("CNN-Dina Mehta.avi" 2008). In the words of Dina Mehta, one of the anchor actors at the time, "There is no leader. . . . It's just everyone pouring in and doing their bit" ("CNN-Dina Mehta.avi" 2008). Actors coordinated hashtags, organized lists of the missing, and provided spaces on blogs to host information about these attacks. According to Mehta (2008a), "We used our influence and experience to get the word out. Our updates and conversations on Twitter kept us connected." Participants, seeming to sense mobilization was occurring, tried to reach out to each other and connect information points for knowledge gaining. This activity is key to knowledge building: Who can help inform? Who can help collect? Who can confirm?

To better understand what happened in Mumbai, we can point to the listing of victims that Mehta and others put together and distributed. The actor-network diagram in figure 5.2 illustrates how the network formed around a central object: a spreadsheet—an anchor actor created on Google Docs. This anchor actor enrolled actors, mobilized them to act, and coordinated this activity. Mehta, a social media specialist, already understood the power of social web tools such as Twitter and Google Docs. Taking up the role of an anchor actor, she drew on her expertise as a social web consultant to rapidly organize her efforts, recruiting volunteers from an extensive network of contacts who then worked to enroll others into the network. Because of her knowledge of social web tools, the network

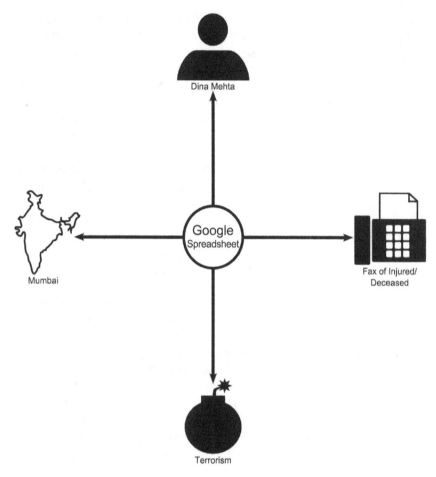

Dina Mehta

Google
Spreadsheet

Mumbai

Fax of Injured/
Deceased

Terrorism

FIGURE 5.2 Actor-network diagram of the initial work of Mehta.

Mehta constructed with others could locate and organize data, cross-reference its information with other sources, and then redistribute the knowledge it had constructed quickly and efficiently (Mehta 2008a). Mehta was thus able to quickly assemble both people and technologies into a robust network that could move through the moments of translation, achieving mobilization of the actors and, finally, punctualization of the network.

During this event, Mehta received a fax listing the injured and deceased that had been recorded at a Mumbai hospital ("CNN-Dina Mehta.avi" 2008; Mehta 2008a), but she could not legally release the fax through mainstream media channels such as television and print news. Realizing the value of the list and the moment of urgency, Mehta turned to the social web for help making the content public and thus enabling the network to mobilize and punctualize. With the faxed list, Mehta created an online directory through which she and volunteers cross-referenced, validated, and dispersed the knowledge they generated. Looking across her available tools, she chose a Google Docs spreadsheet. Using this spreadsheet, volunteers sorted information, alphabetized names, uncovered demographic trends, and searched for specific victims. With such useful tools, Google Docs provides a ready-made system for collaboration during times of disaster.

This Google Docs spreadsheet was named "terror victims—27th Nov 2008" (Mehta 2008b).[3] It contained nearly 300 names and included all available details from the fax and plenty of room for other volunteers to add more details. These details included the name of the hospital where the victim was; the victim's name, age, and sex; and the victim's status as either injured or dead. Each entry also contained the original numbering system from the fax list for marking the victims. Volunteers transcribed parts of the fax into the Google spreadsheet. By adding victim information from the fax to the spreadsheet, volunteers turned data into information, making useful what would have otherwise been static information sitting in someone's closed desktop. Volunteers could then validate information with friends, family, and other sources—including, when feasible, the hospitals themselves.

Writing on Twitter, Mehta recruited volunteers who helped her populate the public Google spreadsheet with data on the injured and deceased (see figure 5.3). Mehta (2008c) referred to herself and another anchor actor in her tweet, asking for help "inputting data on injured/dead." At that moment, social web participants could leverage Twitter's retweet and hashtag capabilities to disseminate information for Mumbai residents and the families and friends of those trapped in the middle of events and to help coordinate efforts for aggregating and organizing information about victims. Through various retweets, participants began to coalesce around the hashtag #mumbai, turning it into a major signpost for sharing information and posting links to relevant information located in other areas. Through Twitter, social web participants began to interact closely with the mainstream media. To coordinate their activities, both groups helped each other share information and raise awareness of the event, distributing information and knowledge (Mehta 2008a).

Through Twitter, Mehta and the many volunteers who stepped forward could mobilize actors within the social web. Twitter's simplicity, combined with its architecture as an open network viewable to those who are not registered participants, supports simple interactions that allow participants to use it as a site for linking together other systems and content. Mehta and volunteers used hashtags to signal the formation and presence of a group within Twitter and then leveraged the system as a way of both coordinating and broadcasting their work. Just as with the use of Craigslist in the context of Hurricane Katrina (see chapter 3), participants appropriated the prescriptions of the system to create a community, in this case, through a stable hashtag. The hashtag, in effect, became an intermediary that supported the inscription of information all through the network. Participants could

FIGURE 5.3 Tweet made by anchor actor during the Mumbai attacks (Mehta 2008c). Tweet by dina; image printed with permission and available at https://twitter.com/dina/statuses/1026464551.

readily share information, recruit other participants, and quickly link to other sites containing information or facilitating further knowledge work.

Moving across multiple social web systems such as Google Docs and Twitter shows a level of literacy and coordination that can help propel mobilization and lead to punctualization. Communities can distribute work, share participation, and validate knowledge. Rather than seeking out a single, perfect solution—often an impossible task anyway—anchor actors such as Mehta coordinated the use of these multiple systems, which resulted in a more robust solution than in prior disaster cases. The *Mumbai Help* blog (2008) hosted numerous postings about this terrorist attack. This blog is a group blog, and those posting to the site functioned as a knowledge repository for the victims and volunteers and others seeking information about their loved ones. It became a curator for this kind of content. One single post garnered numerous responses from people offering to help, asking for help, and trying to distribute information (zigzackly 2008). Thus, the blog was able to enroll other actors, defining the event and identifying the key actors in the space. A link to the Google spreadsheet added to that same *Mumbai Help* post connected Mehta as an anchor actor in both spaces, mobilizing others to validate information and spread knowledge (zigzackly 2008). This mobilization increased the flow of traffic to this knowledge source, thereby transferring it across the actor network.

As we can see in figure 5.4, the actors were spread out across multiple social web tools. These participants employed these technologies in ways that the original producers of these tools never intended. The blog *Mumbai Help* was a space of knowledge curation for the network. While the group was small ("CNN-Dina Mehta.avi" 2008), it was able to accomplish a lot with the tools it had. As Spinuzzi (2008) has pointed out, the mobilization moment creates a space for collective action. Google Docs was the site for participants to organize content. Twitter was the place for people to communicate with each other. With Google Docs as a site to organize disaster content and Twitter as a site for volunteers to organize work, we have the beginnings of a solution for future deployments.

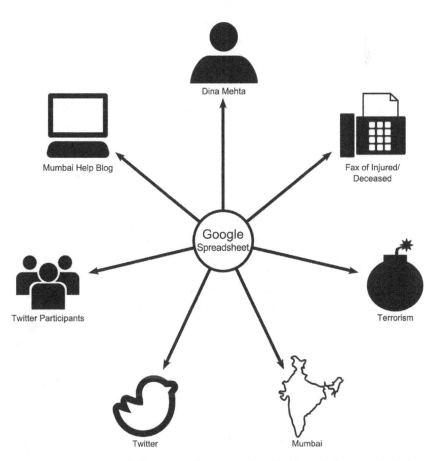

FIGURE 5.4 Actor-network diagram of the actors who aided knowledge transfer during the Mumbai attacks.

Practical Solutions

The Mumbai attacks represented a significant moment for researchers documenting social web literacy, with participants creating punctualizations where none had existed. As participation across these systems reaches tipping points in terms of population and media awareness, people find more ways to communicate and share information. Ethical questions about reporting on missing and found persons certainly exist, but this activity is already taking place. After the Virginia Tech massacre, students exchanged information about the victims on Facebook (Olson 2008). Images on mainstream news sites validated information about victims as well as survivors of the deadly tornadoes in Joplin, Missouri (Martinez 2011). The idea is not for systems to subvert official policies designed to protect the privacy of victims and their families. However, since this work is already taking place, we need to figure out how to create systems that can support validating content in a manner more timely than what traditional, off-line systems do. The work of Mehta and other volunteers who helped her points to the effectiveness of sharing information.

What makes this example remarkable is not actors' possession of this data but rather their digital literacy skills and enthusiasm, matched with the speed and reach of the tools they deployed. Mehta quickly created an ad-hoc social web system, enrolled actors, and had other participants help her verify information. By leveraging an affordance familiar to other computer users, she empowered volunteers. The flexibility of Twitter and Google Docs is to be applauded. Figure 5.5 illustrates this content transformation.

While further validation is always useful, the Mumbai attacks example shows the good that can come of participatory culture. The work of these participants is a signpost to researchers and architects—both of policy and of systems. It indicates these tools' effectiveness when information is not only available on a website but also transferable from one system to another. That said, we still face many challenges related to misinformation, broken systems, and literacy. We need to replicate the effectiveness of Mehta's approach and provide more tools to support this kind of work. The solutions hinge on how experience architects provide ways for people to be active participants, how well the mainstream media cooperate with these participants to further distribute information about key sites of

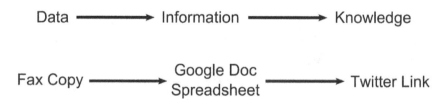

FIGURE 5.5 Content transformation during the Mumbai attacks.

activity, and how empowered anchor actors feel to create their own spaces for communities to meet and exchange details.

Conclusion

Combining the power of human and nonhuman actors is the key to creating better systems to support people in times of disaster. Combing this framework for understanding how content moves through the three stages—data, information, and knowledge—along with an understanding of how actor networks reach translation—through problematization, interessement, enrollment, and mobilization—is one way to construct research projects that can uncover solutions. We are living in a time when natural and human-made disasters are increasing. We also have plenty of everyday communication problems in the workplace and in our social lives that require innovative approaches. Our work is not done here; in fact, the speed that technological, cultural, political, and economic changes occur means our work will never be done.

Mehta's work in response to the Mumbai attacks presents a set of solutions, functional and conceptual, to the communication problems encountered by victims and their families, and we can improve upon that set of solutions. After a disaster occurs, the flow of information becomes critical. However, the closed architecture of systems that handle this information flow presents a barrier, one that Mehta's example proves can and should be torn down in favor of systems that speak to one another in robust ways. Experience architects must begin by asking how participants find groups and content such as tweets, blogs, Flickr photo pools, or collaborative documents. We must build to improve enrollment and mobilization so that volunteers will be able to get their work done with as few problems as possible.

To explore these issues further, chapter 6 turns from ANT to the impact of this research on the future of experience architecture. We need more innovation and more exploration to support the work of our researchers. With this research, we want to support practitioners as they employ techniques to create more usable systems and refine the interfaces that support participants' interactions. We must architect systems that can aggregate information in a way that effectively aids the assemblage of network actors who respond to a disaster. These networks must be able to use the mutable mobiles constructed through social web tools, and researchers must keep in mind the asymmetrical relationship between human agency and the impact of technologies. The technological capacity and cultural interest to improve these experiences clearly exist. Now is the time to discuss how to move forward to find solutions.

6

ARCHITECTING SYSTEMS FOR PARTICIPATION

On the social web, people expect to participate, engage, and interact using a variety of tools within an ecosystem. Unfortunately, many of these ecosystems are unprepared for this level of activity, leaving participants trapped in broken experiences. What can we, under the new banner of experience architecture, do to support participatory culture? How can we help empower people to participate? We need to make a purposeful turn toward working with our participants, engaging them, and respecting their need to engage.

This chapter outlines a pathway for future research and suggests key technologies that could improve the communication problems discussed in this book. For years, the primary focus of academic study in technical communication has been the workplace. We are just now starting to see more research in the social realm, and this wider scope presents us with an enormous opportunity for research. While the social uses of technology and the communication issues of everyday experience typically go unnoticed, social websites of everyday use and participation exemplify how people write and communicate outside large organizations, corporations, and other institutions. Many of us helped make these technologies and have long histories of working on software projects as either researchers or practitioners. Refactoring our roles into experience architecture matters because the shift can help us apply our understandings of form, content, and context to the entire ecosystem.

Certainly, the use of the social web during times of disaster is a largely untapped site of study. But as this book shows, a participatory culture has engaged these technologies, sometimes adding to and often circumventing current system implementations and traditional media channels. This type of participation is not unique to disasters; however, disasters present high-pressure cases that show how well systems can support participation when that participation is urgently needed. As the industry moves toward producing more participatory tools, we must reassess how we can architect for these experiences. Within sociotechnical systems, participants are using multiple tools, data sets, and media, both off-line and online. As experience architects, we must consider the entire spectrum of infrastructure rather than only focusing on issues concerning content creation or information design.

Throughout this book, I address a "we" that is a mix of scholars, practitioners, and teachers who are motivated to make a difference. I am speaking directly to those of us with the know-how to mobilize the strategies heretofore discussed and the opportunities to do so in at least three different arenas: in new technologies by making things, in new research by studying things, and in new courses and programs by making new kinds of professionals. We may have different titles or work in a number of different roles, institutions, industries, and so on, but in this chapter I am calling upon us all to take action. We should work as agents of change for participatory experience.

Looking beyond components that exist primarily at the surface of social web experiences will help us, as scholars and practitioners, reconceptualize technologically mediated experiences. And, in truth, we have been preparing our students to do this kind of work for years now. We spend countless hours in classrooms discussing form and content, discussing visual rhetoric and writing for digital spaces. As a field, we focus on service learning, encouraging our students to work with clients and guiding them through the processes. Our students hold stakeholder interviews, conduct usability testing, and build solutions for clients. We produce graduates who have an appreciation for participants and an understanding that our teachings in rhetoric have real practical applications. They work in project teams, determining strengths—who is best at organizing the group, who can build HTML pages, who is best at editing, and who can do the visual design. Given our strong foundations in these areas, we are prepared to take on the challenge of experience architecture. We are well situated to lead this new movement.

This book aims to fill this gap and address this problem in a way that interests readers within academy and industry. As someone who has one foot in each of those spheres, I think that our work must bridge these spaces if we are to build systems based on experience. We should apply and adapt sociotechnical theories and information science theories to examine participant experiences because doing so allows us to map the uses and movements of these participants. Applying these theories as methods of conducting research in these sociotechnical spaces, we are able to look across the ecosystem of technologies and people to identify and understand the movement of participatory cultures. This approach provides new opportunities for developing tools, policies, and systems to support the work of participants in the social web.

By using this approach, we architects are better prepared to influence the development of these systems and improve participants' experiences. In this way, the use of such an approach relocates humanities-trained workers from the role of "documenter after the fact" to the role of "member of the project team"—a member who participates and contributes as the product is being developed. Using actor-network theory (ANT) and participant-centered practices, we can develop new theoretical frameworks and practical tool sets with which we technical communicators can move beyond our traditional borders and into the architecture of sociotechnical systems. Across these ecosystems, actor networks form by the punctualization of these networks. Following the four moments of translation, actors who help organize the network (anchor actors) begin to define the situation (problematization); actors share these definitions across the network, and we begin to see the network stabilize (interessement). Activity then concentrates, and anchors enroll more actors (enrollment); these actors form the group's assemblage and are mobilized to act (mobilization). By tracing digital culture, we can uncover these movements so that we can improve system products, services, and policies.

Of course, we must become participants and engage other participants as we research and build these systems; otherwise, these new frameworks and methods are useless. The frameworks and methods are also useless unless we integrate these practices into our teaching and mentorship. Examining the moments of translation provides researchers and architects with a close-up view of how connections between actors form during information exchange. Often, actors connect based on some prior familiarity and stability within the adopted site. And yet, in these use cases, it is not always easy to predict which mode of communication gets used and how because the one constant about the internet is that it is always in a state of flux. We have an immense opportunity to take part in the evolution toward making systems more participatory. We need to move beyond single-use systems and build for flexible, contextually situated, participatory systems. As experience architects, we must examine how everyday people extend the information structures and interface paradigms of the systems that we create; by doing so, we can improve system architectures and the policies we create for these ecosystems.

New Disasters: Participant Innovations and Continued Struggles

The following section describes more recent natural and human-made disasters and the response to these disasters across the social web. Here, I briefly trace the actors and networks participating in systems during these disasters. After this brief outline, I discuss participant-centered frameworks and possible solutions. I outline the disasters for several reasons. First, these disaster issues are current and urgent. Unfortunately, the work of tracing disaster is an ongoing process, as new natural disasters and acts of terrorism occur on an almost monthly basis. As I began editing this chapter, the shootings at Sandy Hook Elementary School were taking place, and the use of Twitter, Facebook, and YouTube during this event was astounding. Second, we have not solved the problems outlined in this book. The same issues occur over and over. And while participants certainly gain literacies and new tools aid in this work, we still are not developing participant-centered systems. We have more work to do, work that offers a huge opportunity for scholars and practitioners to align their efforts toward researching and building participatory systems.

Haiti Earthquake of 2010

A devastating earthquake struck the island nation of Haiti on 12 January 2010. With over 200,000 people killed, 600,000 people displaced, and $8 billion worth of damages incurred, this earthquake was as much a disaster as a humanitarian crisis (*New York Times* 2012a). Promising billions of dollars in donations, many people across the world raced to provide Haiti with needed supplies and launch people-finder websites. We can see how different systems reacted to participatory culture in these people finders. The majority of the ones I discuss here failed.

Of all of the spaces where people finders were created after this disaster, the one space where everyday people hoped to find a good user experience was the website of the International Committee of the Red Cross (2010). Solving the issue of CNN's Safe List people finder from Hurricane Katrina, this new people finder allowed visitors to navigate the listings alphabetically rather than numerically. This information design removed the guessing game of wondering whether the last name "Garcon" was on page 5 or 15. Yet even this search function had its faults: the listings were arranged by full name, so, for example, finding Liza Potts meant searching for the entire name "Liza Potts" rather than searching for "Potts" or "Liza." Still, the site did allow everyday people to add content.

During the Haiti crisis, CNN used a list similar to the Hurricane Katrina Safe List. The new list's title, "The Missing, the Found, the Victims" (CNN 2010), was certainly more honest in its assessment of the people on the list, if not more dramatic. Through the iReport section of CNN's website, everyday people could add people to this list. That said, CNN created another walled garden, this time with over 10,000 entries. Users cannot sort the list by name, location, status, updates, age, or any other category, and they will not find an obvious way to export the data. While users can search for people, their searches must be accurate (unlike the forgiving search algorithms on Google). Many entries are incomplete; indeed, one on the front page contains a photo but no name or age. This list is a good start, but the list needs more mechanisms for other people to add details to the original entries. In its defense, CNN is a news source and not a missing-persons finder. It is making improvements, although these improvements are very limited. However, CNN could easily stretch during times of disaster and provide people-finder services. And, as a central repository of information, CNN and other news organizations could easily serve to help unite people in times of need.

If that person finder is what we can expect from a news source, what should we ask of a born-digital technology such as Google? While Google's experiment with its Person Finder for the Haiti earthquake was by far more open—allowing participants to add information—it lacked any anchor actors or moderators from Google itself. Launched by volunteers at Google, this people finder aimed to "prevent the proliferation of multiple missing persons databases" (Fuller and Ramaswami 2010). This goal could easily place Google as the central hub for missing-persons lists for the foreseeable future. While the Person Finder website for the Haiti earthquake is no longer live (Haiticrisis.appspot.com 2012), Google continues to work on a similar project (Google Person Finder 2012). While Google's intentions were noble, the execution of the Haiti earthquake Person Finder was lacking in several ways. First, the tool collected specific data from the participant. Google wanted to know your name (mandatory), your phone number (optional), and your email address (optional). Google included a note at the bottom of the page stating that these data would be made public. For participants who preferred to remain anonymous, the only choices were to enter false information or to opt out of participating. Second, users had no way to flag entries as inappropriate. Scanning through these lists, I quickly lost count of how many false records I found. Third, the archive is now "deactivated and its records have been deleted" (Haiticrisis.appspot.com 2012). Losing access to this information is a major problem for researchers trying to trace past disasters, let alone for those of us who are trying to write these histories. Lastly, and perhaps most important, anchor actors have had little role to play on Google's Person Finder for the Haiti earthquake. Without the ability to flag, what could anchor actors have moderated? Without the ability to edit any of an entry's data, short of submitting a second entry form to update the status, how could they have tried to circulate knowledge? So while Google's Person Finder was a noble attempt, it deeply needs a participant-centered architecture. We need future systems to empower anchors and participants while allowing for technological solutions to help manage content.

New Zealand Earthquakes in 2010 and 2011

In less than one year, New Zealand suffered two major earthquakes. In September 2010 and February 2011, people responded to these earthquakes by using Twitter to exchange information about the event and to assist victims. In particular, the 2010 earthquake was a watershed moment for Twitter use in a disaster. With a magnitude of 7.1, this earthquake was extremely destructive, although thankfully there were no human fatalities (GeoNet 2012). At a cost of around NZ\$3 billion (Bennett 2011), the earthquake and its aftershocks were one of the first major disasters in New Zealand since social media took off there. The internet and mobile devices became a lifeline for these communities without power and with a telephone infrastructure in tatters.

Much like it did during the Mumbai attacks, Twitter became active with tweets about the 2010 earthquake (Potts et al. 2011). Some tweets used keywords to signal the location (Christchurch), while others mentioned the event itself (an earthquake). Rather quickly, multiple hashtags emerged, such as #eqnz, #nzquake, #nz, #quakenz, #christchurch, and others (Seitzinger 2010). By searching for these hashtags, other Twitter participants could follow conversations. Hashtags bring actors together, and during this earthquake, they became an essential way to share information. In less than a half hour, one hashtag emerged more saliently than the others: #eqnz (Seitzinger 2010). Similar to other disaster cases, anchor actors began to organize around this new hashtag and notified participants: "#earthquake Apparently the official hashtag is #eqnz; #doingitwrong. Glad someone's got priorities sorted" (adzebill 2010). Anchor actors also helped others: "thanks megan! RT @harvestbird Try tag #eqnz. Power out in much of Chch, water mains damaged. #nzquake" (tristamsparks 2010), and they organized the actors on Twitter: "@louis_press Note ur tweeting on Christchurch earthquake, fyi official hashtag = #eqnz (not sure who makes it 'official', but . . .)" (morealtitude 2010). The people who make these hashtags "official" are these anchor actors who attempt to restore order to the postdisaster chaos online, structuring the means by which others can exchange information. Unfortunately, they cannot reach everyone on Twitter, so hashtag collisions continued throughout the event.

Japan Earthquake and Tsunami in 2011

In March 2011, an earthquake struck Japan. At a magnitude of 9.0, it was the fourth-largest earthquake in the world and many times stronger than the one that struck New Zealand (NGDC 2012). Similar to people in the New Zealand earthquakes, people in Japan actively used Twitter to connect with victims, families, and supporters. And just like in the cases of New Zealand, problems arose with hashtag collisions (Potts et al. 2011). Encouraging consistent hashtag use across an event is difficult because users must spot these tweets in a sea of thousands. That said, Twitter was acknowledged for its ability to help enable communication in the wake of this disaster (Crump 2011). While Twitter was already an active service, people used it to an extreme level during this disaster. It is, in fact, amazing that it did not buckle under the traffic of 1,200 tweets per minute from Tokyo after the earthquake (Taylor 2011).

Without phone lines to communicate (Crump 2011; Winn 2011), participants in Japan used the internet to share information. Once again, hashtag collisions were a problem. Locating information can be a challenge on Twitter during the best of circumstances, but in a sea of so many tweets, organizing content was difficult. Hashtags such as #japanquake, #tsunami, and others emerged (Farivar 2011). The US ambassador to Japan, John V. Roos (2011), used the more generic #Japan in his tweet trying to connect officials with family members: "If you have friends or loved ones in #Japan that you've been unable to reach, email japane-mergencyusc@state.gov."

Without a way to connect these different hashtags, without a central point at which to gather and then redistribute this content from, Twitter can be an informational black hole if anchor actors do not work to gather, validate, and redistribute this kind of content. We need better systems that can help us organize this information, help anchors locate it, and help stabilize these networks.

Virginia Tech Massacre in 2007, Norway Attacks in 2011

The natural disasters that I have discussed in this book are events that occurred over several days, even if the major catastrophe erupted in one single moment, but mass shootings are often a different case entirely. Natural disasters such as hurricanes and tsunamis tend to result in chaos in the immediate aftermath. In contrast, mass shootings tend to have a much shorter duration between the time people hear about the disaster and the time they find out what happened and to whom. Mass shootings tend to generate different communication trails between the victims and their families as well among their friends, witnesses, and the rest of the world. While first responders may need days or more to locate the victims in bombings and hurricanes, they find the victims of mass shootings far sooner.

But how do people find information online about shootings? Where do they go to find out what is happening, who is affected, and how they can help?[1] As always, the answers to these questions depend on the age of the victims, the location of the incident, and the availability of technology. While victims may be found sooner during shootings, the time between the shootings and the notifying of victims' families nevertheless may leave a vacuum of information. People fill that vacuum using different kinds of social tools. Based on the activity following recent mass shootings, we know that people tend to use text messaging and chat such as the internet-based tools AOL Instant Messenger and Google Talk. While we do see some use of spaces such as Facebook and Twitter, these tend to be used after the event to share grief and news and to memorialize the victims.[2]

In April 2007, a school shooting killed 32 victims and wounded 17 more at Virginia Polytechnic Institute and State University in Blacksburg, Virginia (Virginia Tech Review Panel 2007).[3] Many of the students there communicated with others through chatting tools such as AOL Instant Messenger and text messaging in the aftermath of the event. But Facebook received particular attention. Students used this social networking site to exchange information on their injured, dead, and safe peers (Olson 2008). In these interactions, called a "Chain of Grief," students worked to locate each other and share their grief (Vargas 2007). As Chris Banks stated on Facebook shortly after the event, "We need to get a facebook group started to keep this news story factual and not sensationalized" (quoted on the website of *ABC News* [2007][4]). Other research has also noted that Facebook was a more effective site for communicating than the emergency services

(Bloxham 2008). Palen's research on the use of Facebook and other social web-sites during disasters concluded that "instead of rumour-mongering, we see socially produced accuracy" (quoted in Bloxham 2008).

During disasters and everyday experiences, people use technologies in ways the producers did not intend. People would choose text messaging over voice calls for many reasons, including not wanting to give away their positions to an attacker (*BBC News* 2011). On 22 July 2011, a lone terrorist set off a car bomb in Oslo and went on to massacre students at a summer camp on the Norwegian island of Utøya. Sixty-nine people died on the island, and another eight perished in Oslo (Sollid et al. 2012). During the Utøya massacre, students were able to exchange text messages with their parents. In one case, a young man texted his parents to say he didn't think he'd see them again; thankfully, he survived (Berg 2011). Working together, the victims of the shooting were also able to exchange text messages with one another during the event, warning each other of the location and description of their attacker (Martinovic 2011). While these tools allowed people to exchange messages, they still left parents wondering if their children were safe. We need to consider how to build systems that can help circulate validated information as fast as possible.

Hurricane Sandy in New York and New Jersey in 2012

In October 2012, Hurricane Sandy slammed into New York and New Jersey. Destroying picturesque boardwalks and crippling public transportation systems, this natural disaster killed more than 100 people, left thousands homeless and millions without power, and cost billions in infrastructure damages (*New York Times* 2012b). From the perspective of many participants involved in locating information about the storm, Hurricane Sandy might become best known for its way of showing that "Twitter Is a Truth Machine" (Herrman 2012).

Misinformation can explode across networks during a disaster. Is that a real photo or is it Photoshopped? Are those numbers accurate, or are even more people missing? Sharing content online is fast and relatively cheap compared to the expense of publishing paper newspapers, and the process of reporting has changed dramatically to keep up with the social web. However, as one media reporter noted, "What's interesting isn't that there was fake news—it's how quickly those fakes were exposed and debunked, not just by Twitter users themselves but by an emerging ecosystem of blogs and social networks working together" (Ingram 2012a). In discussing how participants in networks deal with these issues, Spinuzzi (2008) notes that they "continually convince their allies to support them in their aims and form enough alliances that they can work around traitors" (41). Ingram noticed this issue of information validity later in relation to Hurricane Sandy and the Sandy Hook Elementary School shootings: "The way that inaccurate news reports about a mass shooting in Connecticut filtered out through social media has brought up many of the same criticisms as Hurricane

Sandy—that social media isn't an appropriate forum for journalism. But this is simply the way news works now" (Ingram 2012b). We need systems that can help us assess content and distribute accurate information to support this kind of activity.

In an article on BuzzFeed, a website that bills itself as the "First True Social News Organization" (BuzzFeed 2012), the deputy tech editor discussed the process of content transformation that is detailed in this book's case studies:

> Twitter's capacity to spread false information is more than cancelled out by its savage self-correction. In response to thousands of retweets of erroneous Weather Channel and CNN reports that the New York Stock Exchange had been flooded with "three feet" of water, Twitter users, some reporters and many not, were relentless: Photos of the outside of the building, flood-free, were posted. Knowledgeable parties weighed in. (Herrman 2012)

The disasters over the past several years have seen a dramatic increase in the use of social web tools for communication. This relentless sharing of information, this search for knowledge, is one of the greatest attributes of participatory culture. And it is a major reason why we must focus on building participant-centered architectures. Disasters cause very visceral reactions in us. We want to answer all of the questions presented earlier in this book. We want to share. We want to help.

Boston Bombings in 2013

On 15 April 2013, two bombs ripped through the spectator area of the Boston Marathon, an annual race taking place across the Boston area since 1897 (Boston Athletic Association 2013). The bombings occurred at 2:49 p.m. EDT, exploding seconds apart near the finish line of the marathon. Three people were killed and 282 were injured (Kotz 2013). Much of the social web activity surrounding this event involved tracing images in an attempt to locate the bombing suspects. Following this terrorist attack, activity picked up on various social web systems. Everyday people began to share content in an attempt to help spread information about the event and identify possible suspects. In the days after the attack, the FBI asked the public to send in any photos or videos of the marathon and its aftermath (FBI 2013). Participants on sites such as Facebook, Twitter, reddit, 4chan, and imgur were already working on amassing photographs and video. Some even listened to police scanners, signals that were broadcast live over the internet by other social media participants.

While all of these social web systems were active during the event, reddit community members had their data-gathering work publicized in mainstream news reports. Participant oops777 created a new section (a "subreddit") called "Find Boston Bombers" to encourage other reddit participants (known as "redditors") to locate the bombing suspects (oops777 2013). This subreddit garnered much criticism in the media and in the community for wrongly accusing several people of being the bombers. Some other media outlets also published the images of these "suspects."

Within days, the subreddit shut down, and its leaders issued an apology for the "online witch hunt" that had taken place (reddit 2013). This subreddit had overshadowed the productive work of several other subreddits, where participants curated news information and offered aid to victims and first responders (reddit 2013). In particular, the numerous "Live Update" posts provided the FBI's contact information for people who might have "personal info regarding the two suspects" and instructed participants not to post personal information about others online (BostonLiveThread 2013). These threads proved much more useful as a community service.

Through this kind of sharing, people can feel useful to each other and actually become useful to each other. The communication issues outlined earlier must reaffirm our need to become architects of these systems. Our training in technical communication prepares us to do this work. In a recent textbook for undergraduates in technical communication, Tebeaux and Dragga (2010) state, "Every decision you make in developing your document should reflect your audience, their needs, and your purpose" (11). In exchanging the term "document" for "experience," we can begin to consider the entire experience of our participants. We are already doing these tasks for our readers; what I am encouraging us to do is to see the entire spectrum of experience and increase our influence as architects.

Frameworks for Participant-Centered Architectures

Uniting the concepts and furthering the theories of science and technology studies, sociotechnical systems, and technical communication, we can move toward creating flexible systems. Such systems can support people-powered solutions and unite those solutions when necessary, resulting in an improved method for coordinating information and creating knowledge. In positioning my research where these three approaches intersect, my goal is to provide a new and richer method for conceptualizing and implementing the holistic experience. My aim is to describe how these different theoretical perspectives relate to one another and how they might contribute to new ways of thinking about social experience research and architecture, and I have begun to do this through the example of crisis situations. By embedding ourselves within these scenarios, we as scholars and practitioners can understand participants' experience, negotiating often-broken systems and manipulating tools to communicate as effectively as possible. By saying we should embed ourselves, I mean for architects to become a part of these communities, observing how people use these systems by participating rather than by silently observing. By adding tags to images, tracking down missing persons online, commenting on blogs, and witnessing how participants stretch social web tools in ways that the producers of these systems never expected, architects can gain an understanding that simple observation would never provide them because participation demands that practitioners use these systems. And by participating, we will be able to architect systems that people can adopt more quickly and easily when they need such systems most.

A perspective that has grown stale is the idea that any website or software tool exists in some isolated space. All experiences—off-line and online—are connected to a broader ecology. We, as researchers and practitioners, need to experience these ecosystems. As one of the founding members and a past president of the Association of Internet Researchers has stated, "The best work recognizes that the internet is woven into the fabric of the rest of life and seeks to better understand the weaving" (Baym 2006, 86). Social web tools demand a change in approach, a refreshment of our methodologies, and a focus on the agency of social web participants. We need to "focus on being true to lived experience" (Hine 2009, 3). People want to interact and participate with and across systems; they do not want to be trapped in one system, application, or website. We need to research and architect systems that support the flow of information across the social web, not just within a specific segment of it. Supporting the experiences of actors is the key to creating more contextual and open mediated systems to support people in their everyday tasks as well as in high-stakes disasters.

We are in need of frameworks that can allow us to improve how we architect experiences, how we build for participation, and how we can do so based on evidence from observing, experimenting, and participating. These frameworks must emphasize participation. In looking across systems rather than narrowing our focus to just a single task, we can build for experience. Basing our approaches on—and evolving further—our research on actor networks and user-centered design, we can refocus our efforts on building participatory systems that invite this kind of activity rather than deter it.

This need for participation includes scholars and architects as well as our educators and mentors. Those of us who are responsible for teaching and guiding students and junior colleagues are ethically and politically obligated to equip future and current architects, engineers, writers, designers, and policy makers with methods, skill sets, and knowledge to research and build for participatory culture. Examining the critical issue of experience from the perspective of participants allows us to answer the call for a more holistic view of communication with appropriate methods to address these situations (Sullivan 1990).

As stated elsewhere in this book, researchers and practitioners—anyone who impacts the development of sociotechnical systems or teaches those who will impact them in the future—are fundamentally obligated to address the dilemmas detailed throughout the previous chapters. Such moves must inform the architecture of the future systems. The social web itself demands a new set of paradigms for understanding the experiences of people who participate in spaces such as Flickr, Twitter, Craigslist, Facebook, and any of the myriad of new social web systems that seem to appear and disappear on a weekly basis. A time of disaster heightens people's purpose, emotional resonance, and activity. Such an atmosphere can lead to developing new paradigms. These paradigms can mean the difference between reuniting family members and leaving them scattered across both digital and physical spaces, between connecting with loved ones and missing their appearance on a list of missing persons, between literal life and death. To do this kind of work requires resituating what it means to be a researcher and a communicator and embracing the idea that "software is a social exercise" (Ratcliffe and McNeill 2012, 25).

Becoming Participant-Researchers and Architects

So how do we build participant-centered experiences? Savage (2004) states that "technical communicators often feel like outsiders in the specialized technology cultures in which they work" (181). Is this a sentiment that arises out of our work or out of the way we present ourselves professionally? We have spent years writing about problems that users have with products—problems that could be solved if our products were more user centered and, now, participant centered. Savage goes on to say that this "marginality may need to be understood as a necessary dimension of practice in our field" (181). I agree that the marginality is a strength, but this marginality is no longer on the margins. By this I mean that now both software industries and higher education institutions are finally at a moment of understanding the need to develop systems that are user centered. We are the ones who are best able to work in spaces where "not mastering but negotiating continually shifting technologies, institutions, discourses, and cultures" is necessary (189).

Because of our preparation as humanists, technologists, and advocates, we are responsible for leading these discussions. Having worked in software development and having watched changes in industry over the last 10 years, I can firmly state that we have a major opportunity to be at the center of these conversations. As Savage (2004) says, the "technical communicator-rhetorician, like the trickster, is an agent of social change" (183). We are these communicator-rhetoricians. Now is our moment to become pivotal contributors and leaders. We must become participant-researchers and participant-architects. We must experience these spaces firsthand. We can no longer collect data without knowing what we are collecting, without understanding how these systems work, without understanding how and why participants are using these tools. We can no longer build new systems or improve existing ones without having accounts in these systems, without participating in them regularly, without coming into contact with the other systems that our participants use. The days of the academic standing on the sidelines are over. The days of the practitioners holding their customers at arm's length are done. If we do not make these moves, our colleagues and competitors will surely leave us behind. And our participants are already doing so. We must join them and be the agents of social change.

An experience researcher and architect has to be within an event, sharing the experience of the participants and seeing firsthand the collected and exchanged. As in contextual inquiry, researchers and practitioners need to place themselves in these spaces to better understand communication activities (Beyer and Holtzblatt 1998; Courage and Baxter 2005). In the words of one of the leading internet anthropologists, "The only real way to learn to be a researcher is by experience" (Hine 2005, 2). My aim in this book is to encourage architects to embed themselves within these participant experiences, whether the architects are building software used by medical personnel, services experienced by retail customers, or processes used by office workers. And in the case of the social web, we can certainly join and engage in Twitter, Facebook, Instagram, and a host of other systems that participants experience in these ecosystems.

As Halvorson and Rach stated in their recent book on content strategy, "Although many of us truly believe we know exactly what our end users *really* want from us online, we can't know unless we ask them" (2012, 10; italics in the original). To become a participant is to go one step further. This work is imperative if we are to be responsible researchers and practitioners. Before practice, we must research. Whether that research is in the academy or in industry, we need to understand before we can begin to solve. Therefore, we must participate. This method takes us beyond the "researcher as lurker" (Rutter and Smith 2005, 87). In this way, the participant-researcher is positioned to identify and experience closed systems. Trapping ourselves in these walled gardens while experiencing other technologies that allow for greater information flow can aid in understanding the need and want to share information. Through this lens, we can see the benefits of tracing participant activities and locating information within events rather than simply creating social web tools from the outskirts of these interactions. We can sketch networks and create maps that can become talking points for research teams and stakeholders, "mapping as a method of inquiry and knowledge creation" (Dodge 2005, 113). This departure from traditional single-experience design to architecting for experiences across ecosystems is imperative if we are to create systems that can support communication between disparate technologies and separated people during times of disaster and for everyday purposes.

Engaging Participants as Co-researchers and Co-architects

How can we engage with participants on the social web? As we move toward becoming participant-researchers and participant-practitioners, we also must view participants as co-researchers and co-architects. As Savage (2004) states, this work "seems to be consistent with a sophistic practice in which knowledge is always contingent, in which rhetorical purpose must be reconciled to the needs of a particular audience at a particular time and place" (189). Above all, this is my major argument.

We must research with participants rather than be passive observers commenting on their work. We must engage them in building and modifying these systems. They deserve a seat at the table, a pen at the whiteboard, and access in ways that allow them to work across these ecosystems. We can and must engage with these participants, or they will engage with more forward-thinking researchers and architects.

In engaging participants as researchers, we must move away from simply reporting on our observations. By engaging participants and asking them to be co-researchers, we benefit greatly from their perspectives and insights. Ethical issues will arise, surely, such as when to publish certain information that could endanger these participants (Association of Internet Researchers 2012). However, given the right circumstances, researchers can partner with participants and publish together. For example, my graduate students and I collaborated with participant Joyce Seitzinger in the New Zealand earthquake research (Potts et al. 2011). Adding her to the project gave a rich texture to our research that would be woefully absent without her voice.

We must move away from the earlier concept of placing the designer within the heroic narrative of saving the user from the pitfalls of interface designs (Johnson 1998). In a way, these new moves can be similar to participatory design like contextual inquiry is (Spinuzzi 2005). Here, I am suggesting that we do these activities within the context of social web spaces rather than in labs or design studios. I am calling for architects and researchers to critically attend to these participants by engaging them within these digital spaces where participants use available tools. We can do so by emphasizing "engagement with the technology and richer collaborations among" our participants (Zappen and Geisler 2009, 3). I am also calling for us to give participants more access to tools for distributing content and accessing data, such as through application programming interfaces (APIs), mash-ups, and other third-party applications (Potts and Jones 2011). By engaging people within these structures and participating in these acts of exchange, we can understand participants' work-arounds. Working with these participants, researchers and practitioners can then explore how these tools may evolve to support experience rather than hinder it.

Participatory Futures

Our futures are participatory. People now walk around with more computing power in their pockets than ever before. We have mobile devices that can use augmented reality to help us find locations boosted by sophisticated GPS systems. Through various open-source projects, we have access to suites of software that enable us to edit movies, tweak images, add blog comments, and distribute videos. And participants' need to share their experiences and connect with others is palpable. In their book on networks, Rainie and Wellman (2012) paraphrase sociologist Ronald Breiger when they say, "People link groups, but groups also link people" (202). Via such links, these experiences occur. In debating the merits of sharing, agency, and access and balancing revenues with openness, the architects of these systems have had their hands full racing to create these structures and attract attention. Politically and economically, we are living in a very tumultuous time where many governments and corporations are in upheaval, and some of this work is playing out across the social web (El-Nawawy and Khamis 2012). The lid is off. The need for future research and exploration is evident.

While some of us may support already-established communities, many of us are researching and building for temporary networks. Especially in the case of disaster, people use technology to connect, complete a task, and get out. In comparison, user-centered design generally views such scenarios and tasks as repetitive in nature, though some can be considered one-time tasks that are rarely repeated.[5] How do we instead support tasks that are less repetitive—ones that require more flexible systems? To build for experience, we need to understand the minutiae of information exchange. These exchanges are instances of connection that are made possible because elements are not inherently related to one another. In these changes, participants can reconfigure connections quickly and effectively to perform tasks. Allowing participants to post images on Flickr, share them through links on Twitter, and edit them in iPhoto is an example of how multiple systems can adapt to provide a space for such participation. Participating directly in these systems, communities, and moments of use makes experience architects intimately aware of how these connections are made, maintained, and severed, as well as why. The case studies that I have presented trace these exchanges for exactly these reasons. We can learn how people and technologies interact with each other and how people are interacting through technologies, and in so doing, we develop a better picture of how mediated systems help or hinder the activities that people pursue.

Many practitioners and researchers working in user-centered design have read and refer back to certain foundational works that draw together the ideas of design and usability. One of those foundational pieces, Norman's (1989) *The Design of Everyday Things,* has a particular passage that is key to the research on the ease-of-use of mediated systems. Norman discusses how everyday activities such as brushing teeth, getting to work, and meeting with friends need to be shallow and narrow. Activities that are not everyday tasks include more complex activities such as playing chess or completing a crossword puzzle. These are activities that are "devised so as to occupy the mind" on purpose and, therefore, they give those who pursue them a challenge (Norman 1989, 124). The structures of these complex activities are deep and wide. Many websites operate in this same manner: they are "sticky" and keep the user's attention so as to generate advertisement page views, brand loyalty, or increased revenues for the website. However, Norman suggests that these are rarely the user's goals; instead, users are there to accomplish some task—to get in and get out.

Unfortunately, our current systems are rife with what Norman would consider to be deep and dangerous crevices and wide and diffuse experiences. Depending on the topic, these crevices and experiences can be useful, but we also need ways for participants to be able to access content in the shallow and narrow spaces. How can we surface useful content? Many of these mainstream systems often lack open-review systems—such as those that are found in social web tools—where participants can improve upon content, language, and sometimes even design structure. Worse yet, many of these systems and the experiences they foster lead to information overload. Participants encounter so much content that they are not certain which is most relevant to their context. In similar readings in Norman's era, we are told a piece of software is like a house, and we should not shuttle users from room to room (Cooper 1995). This tenet of design is a good one: do not drag people through unnecessary hoops as they accomplish their tasks. Applying this concept to architecting social web tools is important; asking our participants to jump to multiple sites and interactions to create participatory content that is compelling and useful to a community is not productive.

Turning to works specifically aimed at technical communicators, a number of publications are relevant to this discussion. In the realm of textbooks, Kimball and Hawkins's (2004) *Document Design: A Guide for Technical Communicators* does a particularly good job of integrating theory and practice so that one informs the other in a way that is incredibly useful for preparing future experience architects. In describing information design, the authors state that it "deals with the relationships among people who create the information, people who use the information, and people's cultures, societies, and environment" (Kimball and Hawkins 2004, 3). Of course, the same can be said for interaction design, information architecture, and user experience. What makes our training different is our focus on human experience. "Audience," "purpose," "form," "content"—these are all terms we regularly use in our classrooms when we discuss how best to consider rhetorical situations and create responses to them. This focus is missing from too many digital experiences. Given our background in rhetoric, our training in technologies, and our experiences with practical application, we are well situated to produce a new generation of scholars and practitioners who can take their rightful place as experience architects on development, policy, and executive teams.

One of the many challenges of researching technologically mediated communication is the speed with which technologies evolve, changing the ways we handle data, implement systems, design interactions, and organize information. However, some aspects stay constant. Halvorson and Rach (2012) lay out several concepts for emphasis. Some of their chapter titles make for good areas of focus: "Problem," "Solution," "Alignment," "Analysis," "Content," "People," "Persuasion," and "Advocacy " (viii–ix). Such focus needs to spread beyond content and into experience. Given our emphasis on and understanding of these areas, our opportunity is clear. As experience architects, we can be the agents of change for products, services, and policies in digital spaces.

Frameworks for Exchange

To examine these experience issues further, we can turn the discussion of pre-scription and inscription. Prescription is "what a device allows or forbids from the actors" (Akrich and Latour 1992, 261), and inscription refers to artifacts that actors create as part of their knowledge work (Latour [1987] 2003, 218). Across the social web, producers create prescriptions that must be followed by partici-pants. These prescriptions are often most obvious in interfaces where people work to share content. Examples of prescriptions include functions such as add-ing tags in Flickr or adding comments in blogs. Participants leave inscriptions in these spaces, signaling their use of these tools. Inscriptions can include leaving tags, uploading images, and posting comments and other people-powered con-tent. These two key components will give insights into how the interface con-ventions and system structures bind participatory experiences and will illustrate how participants are turning raw data into actionable information.

Architects need to pay attention to the ways in which these participants inscribe their experiences in these systems. Discussing the term "network," Latour ([1999a] 2005) states that the concept is "the *summing up* of interactions through various kinds of devices, inscriptions, forms and formulae, into a very local, very practical, very tiny locus" (17; italics in the original). Simply building prescrip-tions, or rules, for a system and walking away is not enough—especially on the social web, where individual sites exist in ecosystems and networks of networks. In response to a given situation—particularly a disaster affecting thousands of people—more than one actor network will operate. Social web participants often work across networks or find themselves working in more than one network at a time in different contexts. Sometimes they are aware of each other, sometimes not. Aside from social connections among participants, the central aspect of the social web is this mobility of content. Participants copy and repurpose content with frequency and speed. Experience, then, is not simply a question of techno-logical use, but of social use and the value of content to participants. Acknowl-edging this breadth is architecting for experience.

The assemblage of actors into an operational network rests at the intersec-tion of these experience concerns and the capabilities of technologies that either support or hinder the efforts of participants. The images posted to Flickr, status updates posted to Facebook groups, tweets on Twitter, collaborative spreadsheets created in Google Docs—all of these are instances in which we can measure the effectiveness of communication by the mobility of content. We can also measure this effectiveness in how that mobility contributes to the work of participants as they cross-reference and validate content and then mobilize in response to their findings. Translation cannot occur without mobility. As such, "mobilization is a way of accomplishing a collective solution, a collective representation" (Spinuzzi 2008, 90). Unless social web content can move effectively and intelligently, pre-scription does not help the inscription of meaning, and neither mobilization nor punctualization can take place. This network mobilization rests on how well the system's prescriptions are built and how easily those inscriptions can be left behind.

In the case of Flickr use, participants operating in or connected through networks apply their own inscriptions to images by attributing semantic values to the images, thereby turning the photos into intermediaries that can then help in constructing the narrative for an event (Latour [1987] 2003). By inscribing this information, participants repurpose it for others within that culture, validating the importance of those images to people affected by that particular event. By taking these pictures and giving significance to them, participants are able to translate material for their groups, leading to punctualization.

In the wake of disaster, both mainstream media sites and social websites attempt to inform victims and their families and collect information. Generating knowledge through the validation of data is critical for creating and maintaining credibility in both spheres. During emergencies, participants tend to go to familiar sites where they have existing social networks. Unfortunately, these sites are rarely equipped for the demands of victims and their loved ones. For researchers and practitioners, then, the scope of design and usability problems extends far beyond simply understanding the use of multiple online spaces and into the effects of physical space and limitations that arise from the impact of the disaster itself. A disaster may severely damage communication infrastructures and limit access to social web tools. It may displace victims not only from their homes, but also from their cities and towns. Coupled with the fact that users must find and reconnect fragmented information across the web, lack of access presents a significant challenge in architecting systems that perform the tasks of translation more effectively and efficiently. We must architect these experiences in ways that allow for actors to flow across these networks.

Participants in the social web are already implicitly and explicitly pointing out the means by which the collective intelligence of online communities can overcome this challenge. Lévy (1997), for example, theorized that such communities can add the individual efforts of their members together and leverage their collective efforts to achieve more than might be thought possible. Moreover, Jenkins's (2006) work on participatory cultures shows the power of the collective intelligence of online communities in tracking down information related to political protests, television programs, and other media experiences. The examples traced in this chapter and the previous chapters illustrate similar principles in how participants use the social web in response to disasters and crises. Collective intelligence—the ability of a community to pool knowledge resources and perform massive amounts of knowledge work quickly and effectively—is one aspect of punctualization. But systems that support the inscriptions within the network must aid this work. Such systems will allow participants to either adapt systems' prescriptions more readily or alter those prescriptions all together.

Social Web Tools as Contextual, Flexible, Responsive

As our research methods improve so that we can better communicate across teams and as our frameworks improve to meet the demands of building social web tools, our social web tools will also improve. Research and product teams will be able to refocus their efforts on experience. By employing the methods in this book, we can discover participant needs and participatory contexts. Researchers and practitioners focused on participants can lead research and development processes, building stronger insights for architecting participant-centered experiences.

In describing these actor networks, Callon ([1999] 2005) states:

> The identity of the actor and the action depends precisely on these configurations, and each of them can be understood only if we agree to give humans all the non-humans which extend their action. It is precisely because human action is not only human but also unfolds, is delegated and is formatted in networks with multiple configurations, that the diversity of the action and of the actors is possible. (194)

By engaging in discussions and projects typically thought of as residing in the realm of software developers and visual designers, experience architects can be stronger advocates for those participating in the social web. In doing so, we can recognize that social web tools must be contextual, flexible, and responsive.

The social web is contextual. In *Ambient Findability: What We Find Changes Who We Become,* Morville provides several meanings for "findability":

> The quality of being locatable or navigable.
> The degree to which a particular object is easy to discover or locate.
> The degree to which a system or environment supports navigation and retrieval. (Morville 2005, 4)

Each of these statements points to issues of experience architectures both above and below the interface, within data structures, and across multiple systems. The findability of information is, then, also central to social web experiences. This centrality means that systems must be able to speak to one another meaningfully and thus aid participants as they communicate with one another, too. Whether architecting systems aimed at workplace communication or creating social web applications that support web-based communication with family and friends, researchers and practitioners must support the knowledge work of participants operating within social environments that value the gathering and repurposing of content. Doing so is especially vital in moments of disaster, times in which these exchanges become critically important.

The social web is flexible. Fischer (2011) argues that practitioners must "underdesign for emergent behavior." Architects must work to create "seeds for open, living information repositories and contexts in which participants can create content, cope with exceptions, design work-arounds, and engage in negotiations" (52). While we are always struggling to balance new features with maintaining current deployments, we must continue to resist the urge to overdesign these systems. Social web use during times of disaster has shown us that people flock to where their networks are—they resist new systems and instead look for familiar spaces. Those spaces need to be flexible so that participants can inscribe new kinds of use on them.

The social web is responsive. Baym (2010) examines issues of communication from the standpoint of mediations, social cues, networks, and issues of authenticity. Delving into these concepts is crucial for designing viable technologically mediated systems for the social web and particularly for those individuals and groups responding to disaster and crisis situations. Across the social web, actors form assemblages (or series of networked relationships) that might exist only for short times in order to respond to specific events. We need to build systems and policies that can respond to these needs in ways that are culturally situated and authentic. We should build for what Halavais (2009) refers to as "collaborative filtering sites," systems that provide answers for participants "in a way that requires little action" from these participants (163).

These are key points that will allow researchers to focus on content and context while investigating the interactions between participants online. In preparing future technical communicators to tackle these issues, we can look to focusing curriculum on experience architecture. Building on our strong foundations in information design, visual rhetoric, and analysis, we can expand these programs to include courses where this focus can be the central discussion. In such courses, we can teach the fundamentals of research in regard to participant observation, stakeholder interviews, and usability testing. We can teach our students about strategies for presenting content as well as structuring that content within systems. We can discuss intellectual property as well as participant access to digital content.

Most of all, we need to emphasize students' need to lead these projects from the perspective of an experience architect. Project teams need leaders, whether these teams are research focused, product focused, or policy focused. But each role in a team is critical for building these experiences, and project leadership is not necessarily the focus of design, development, or simply managing a project. Many of us who will do this kind of work come from different backgrounds, institutions, and companies. We may not always share the same titles, but we need to share the same focus on participants and experiences. Focusing on participants is our responsibility as advocates who have the skills to look across these ecosystems. Carolyn R. Miller (1979) gave us her "Humanistic Rationale for Technical Writing." Let this book stand as our humanistic rationale for experience architecture.

Conclusion

To paraphrase Law (2004), our world is complex and inherently messy. Researching the social web is deeply messy, and practitioner solutions are not always readily at hand. This book is just the beginning of the conversation on how to research and model systems to encourage, empower, and bridge participatory cultures on the social web. We are quickly moving past design work that looks at individual tasks and interfaces based on one user in one system. We need to design for participants who span several systems as they move from activity to activity. And their request is simple: "Let me act" (Mehta 2008a).

We need experience architects who are participant focused and can work across project teams with the goal of supporting actors across these ecosystems. As Laurel (2003) states, "Design has power. . . . Design has consequences" (19). We must stay focused because "the internet will often not be experienced as a single entity and will have many different social meanings" (Hine 2009, 5). We must embed ourselves, and we must participate. Employing ANT in this book, I have argued that we must shift our perspectives from a user-centered viewpoint to a participant-centered one. As act evolves into activity and reception evolves into production, we must move toward architecting for *how* these activities can take place, given the network of potentially traceable associations enabled by networked structures. We can architect experiences to support people locating data, validating information, and exchanging knowledge. This resulting methodology is based on the notion that people, technologies, and organizations are all part of these ecosystems of experience.

Tracing these temporary connections is invaluable to understanding how people accomplish these activities. As Baym (2006) states, this work "has been and continues to be essential in shaping our understanding of the internet, its impact on culture, and culture's impact on the internet" (79). Such understanding can lead to improved systems, processes, and interfaces. Embedding architects as participants within communities of practice can enable us to build for scenarios that we ourselves have experienced. Within a product team or a research group, the diagrams describing these activities can lead to a shared knowledge of the relationships between people and technologies. With that knowledge, these teams can better design the information architectures, user interfaces, system frameworks, and policies.

In recognizing these instances of connectedness, researchers can begin to map who these actors are (e.g., mobile-application participants), where they are participating (e.g., from the comfort of their homes, from the middle of disasters), and what events are affecting their participation (e.g., slowed cell phone reception). Tracing the actors within these participatory experiences can, for example, help researchers and practitioners visualize system collisions and communication misfires and pinpoint usability issues. What triggers people to participate in events online? How do they locate specific websites to validate content? What are the specific moves they make in order to update information?

To begin to answer these questions, researchers and architects need to observe how participants are building knowledge within these systems as well as participate in those spaces themselves. Though control of experience is one objective for this interplay of forces, these systems also empower subjects who join together and work through systems to accomplish their goals. In the case of disaster, participants locate, validate, and distribute knowledge across their communities; they don't wait for officially sanctioned activities. Examining these details within these experiences will help researchers and practitioners build systems that can meet the communication needs of these social web participants. While "designers cannot predetermine and prescribe users' actions anymore than users can apply a particular piece of technology exactly as they like" (Bødker and Iversen 2002, 12), through these actor-network maps, experience architects can begin to understand how participants are transferring knowledge across these systems. To do so, however, requires an understanding of the most fundamental of processes involved—becoming part of this activity in order to understand this activity.

We cannot ignore the social web. We have only just begun to scratch the surface of what social web experiences can be. Lévy (1997) has stated that the internet "could become the most perfectly integrated medium within a community for problem analysis, group discussion, the development of an awareness of complex processes, collective decision-making, and evaluation" (59). Researchers and practitioners must see these systems as dynamic, fluid spaces in which the flow of information is a key goal. Participants in the social web are not satisfied with static spaces in which the implementations channel them through tightly regulated social interactions or dictate what their social behavior should be. Instead, activities such as tagging, labeling, organizing, and moving content are fundamental to the social experiences participants now demand. Participants want to communicate, but they must be able to find relevant content and one another. Systems that balance the structured and the unstructured, the rigid and the flexible, are the ones that have the most chance of success. Not only do architects learn from actor networks by participating in actors' work, the systems we create can do so as well. The knowledge we researchers and practitioners gain from our experiences as participants can inform how we architect and develop such systems. For educators and mentors, such experience can significantly help train the next generation of researchers and practitioners. Social systems—both online and off-line—continue to evolve. We have a huge opportunity, as well as an enormous responsibility, to be part of that evolution in ways that can lead to more contextually aware experiences for our participants. Let's get to it.

APPENDIX: THE NOUN PROJECT ICONS

The following is a list of the icons from The Noun Project that I used throughout this book. Although I could have designed stencils in Illustrator as I have for past projects, I wanted to include the work of these artists as part of my overall argument for making our work more participatory and inclusive. I have cited each icon according to The Noun Project's requirements found here: http://thenounproject.com/using-symbols/. I am grateful to these artists and to The Noun Project itself:

Bird designed by Thomas Le Bas from The Noun Project.
Bomb designed by Adam M. Mullin from The Noun Project.
Building designed by Nate Eul from The Noun Project.
Calendar designed by Marcio Duarte from The Noun Project.
Camera Phone designed by Roy Milton from The Noun Project.
Circle designed by Thomas Le Bas from The Noun Project.
Coastal designed by Iconathon from The Noun Project.
Community designed by T. Weber from The Noun Project.
Desk designed by James Thoburn from The Noun Project.
Explosion designed by Renee Ramsey-Passmore from The Noun Project.
Fax Machine designed by Braden Stranks from The Noun Project.
Gears designed by Dima Yagnyuk from The Noun Project.
Hash designed by P. J. Onori from The Noun Project.
Hurricane designed by The Noun Project.
India designed by Satheesh CK from The Noun Project.
Information Technology designed by United Nations OCHA from The Noun Project.

London Underground designed by Viktor Hertz from The Noun Project.
Network designed by The Noun Project.
Notebook designed by Brendan Lynch from The Noun Project.
Picture designed by mooooyai from The Noun Project.
United States designed by James Keuning from The Noun Project.
User designed by Denis Chenu from The Noun Project.

NOTES

Note to Chapter 1

1. Bryson (2001) notes this disconnect when he relates a story of having to take multiple trains to move between two Underground stops that are physically located across the street from each other.

Notes to Chapter 2

1. For a visual description of a distributed network, take a look at the diagram on the Rand Corporation (2011) website: http://www.rand.org/about/history/baran.html.
2. For a more nuanced discussion of equal agency, or symmetry, in ANT, I highly recommend a series of discussions on Clay Spinuzzi's (2012a, 2012b) blog. The first discussion in the series is found here: http://spinuzzi.blogspot.com/2012/03/symmetry-as-methodological-move-part-i.html.

Notes to Chapter 3

1. On 26 December 2004, a devastating 9.1 undersea earthquake caused multiple tsunamis that affected numerous countries across Asia. With over 229,866 persons missing or dead (United Nations Office of the Special Envoy for Tsunami Recovery 2010), this disaster serves as an early example of the confusion generated by the collision between the social web and traditional one-to-many websites such as those of *BBC News* and the Red Cross. In the aftermath of this Indian Ocean earthquake, participants rushed online to help find the missing and locate information on injured relatives, view photos, and donate to various nongovernmental organizations (NGOs). These moments provided a high-pressure environment that tested the usability and appropriateness of the various tools employed by these participants.
2. Numerous examples of unsavory uses of Craigslist in the wake of Hurricane Katrina exist (Axline 2005).

Note to Chapter 4

1. David Storey has since changed his username on Flickr. This kind of change is a hallmark of the internet and makes using the internet in research particularly challenging.

Notes to Chapter 5

1. Within the social web, information includes every form and mode of media available through computer software and internet communication technologies: text, images, videos, and sound, to name but a few. Moreover, it can include the different types of metadata and scripted actions embedded into content, including links and interactions.
2. Friedman (2005) addresses this concept in *The World Is Flat: A Brief History of the Twenty-First Century.*
3. I am not including a screenshot of this Google Docs spreadsheet out of respect for the victims and their families. As of this writing, the spreadsheet is still available online for viewing.

Notes to Chapter 6

1. Locating this information during crisis situations is difficult at best. And after finding the information, we have other difficulties in curating it. To give a sense of how search engines handle these issues, I use the example of those trying to search in Google for the Virginia Tech mascot. "Virginia Tech massacre" appears in the autocomplete feature. While autocomplete is a useful feature in Google, in this instance it is socially awkward at best and deeply offensive at worst. Regardless, this example shows that our tools need to better understand context. They must. And we must devote more time to this kind of research and practice.
2. During the final editing of this chapter, the 2012 school shooting at Sandy Hook Elementary School was unfolding. As of this writing, details are still emerging on the victims, and data are being gathered about the use of social web tools.
3. Deciding whether to include the perpetrators in these casualty counts is a topic of debate among participants using social tools. Most of the contributors tend to agree that victims should be listed separately from their murderers ("Talk: Sandy Hook Elementary School Shooting" 2012).
4. The Virginia Tech massacre is also considered to have marked a major shift for journalism. An interview of reporters investigating this disaster discusses this shift in more detail (Thompson 2010).
5. An example of such a one-time or "one-off" task is requiring users to set up account information when they sign up on a banking website. After the users create an account, this activity is not one that they would most likely revisit.

REFERENCES

ABC News. "If You're OK, Please Update Your Profile." *ABC News,* 16 April 2007, http://abcnews.go.com/US/story?id=3046253&page=1#.UM3jU7ZkhQA.

adzebill. (2010). Twitter / @Mike Dickison: #earthquake. Apparently the official . . . Twitter. Retrieved from https://twitter.com/adzebill/status/22905414859.

Akrich, Madeleine, and Bruno Latour. "A Summary of a Convenient Vocabulary for the Semiotics of Human and Nonhuman Assemblies." In *Shaping Technology / Building Society: Studies in Sociotechnical Change,* edited by Weibe E. Bijker and J. Law, 259–64. Cambridge, MA: MIT Press, 1992.

Applegate, Chris. (2005). London tube bombing. Flickr. Retrieved from http://www.flickr.com/photos/qwghlm/24230239.

Association of Internet Researchers. "Ethics Guide." Association of Internet Researchers. Accessed 30 December 2012. aoir.org/documents/ethics-guide/.

Axline, Keith. "Craigslist versus Katrina." *Wired,* 1 September 2005. http://www.wired.com/medtech/health/news/2005/09/68720.

Baym, Nancy K. *Tune In, Log On: Soaps, Fandom, and Online Community.* Thousand Oaks, CA: Sage, 2000.

———. "Finding the Quality in Qualitative Research." In *Critical Cyber-Culture Studies,* edited by David Silver and Adrienne Massanari, 79–87. New York: New York University Press, 2006.

———. *Personal Connections in the Digital Age.* Malden, MA: Polity, 2010.

BBC News. "Eyewitness: Norway Utoeya Shootings." *BBC News,* 22 July 2011, http://www.bbc.co.uk/news/world-europe-14255004.

———. "7 July Bombings: What Happened; Overview." *BBC News,* accessed 28 December 2012, http://news.bbc.co.uk/2/shared/spl/hi/uk/05/london_blasts/what_happened/html/.

Bennett, Adam. "EQC 'Could Deal with Another Big One.'" *New Zealand Herald,* 2 March 2011, http://www.nzherald.co.nz/business/news/article.cfm?c_id=3&objectid=10709579.

Berg, Ulf. "Utøya Shootings: 'Our Son Texted Us—I Love You. I Don't Think I'll See You Again.'" *Observer,* 30 July 2011. http://www.guardian.co.uk/world/2011/jul/31/utoya-shootings-ulf-berg-son.

Beyer, Hugh, and Karen Holtzblatt. *Contextual Design: Defining Customer-Centered Systems.* San Francisco: Morgan Kaufmann, 1998.

Blakemore, Michael, and Roger Longhorn. "Communicating Information about the World Trade Center Disaster: Ripples, Reverberations, and Repercussions," *First Monday* 6, no. 12 (2001), http://firstmonday.org/htbin/cgiwrap/bin/ojs/index.php/fm/article/view/907/816.

Bloxham, Andy. "Facebook 'More Effective Than Emergency Services in a Disaster.'" *Telegraph,* 30 April 2008, http://www.telegraph.co.uk/news/1914750/Facebook-more-effective-than-emergency-services-in-a-disaster.html.

Bødker, Susanne, Pelle Ehn, John Kammersgaard, Morten Kyng, and Yngve Sundblad. "A Utopian Experience: On Design of Powerful Computer-Based Tools for Skilled Graphical Workers." In *Computers and Democracy: A Scandinavian Challenge,* edited by Gro Bjerknes, Pelle Ehn, and Morten Kyng, 251–78. Brookfield, VT: Gower, 1987.

Bødker, Susanne, and Ole Sejer Iversen. "Staging a Professional Participatory Design Practice: Moving PD Beyond the Initial Fascination of User Involvement," in *Proceedings from NordiCHI '02,* 2002, 11–18.

Boston Athletic Association. "History of the Marathon." BAA. Accessed 19 April 2013. Retrieved from http://www.baa.org/races/boston-marathon/boston-marathon-history.aspx.

BostonLiveThread. Boston Marathon Explosion—Live Update Thread #17. Posted to the inthenews section of reddit. Accessed 19 April 2013. Retrieved from http://www.reddit.com/r/inthenews/comments/1cn67p/boston_marathon_explosion_live_update_thread_17/.

boyd, danah. "Why Youth (Heart) Social Network Sites: The Role of Networked Publics in Teenage Social Life." In *Youth, Identity, and Digital Media Volume,* edited by David Buckingham, 119–42. Cambridge, MA: MIT Press, 2007.

Bradshaw, Tim. (2005). Tavistock place. Flickr. Retrieved from http://www.flickr.com/photos/timbradshaw/24284534/in/pool-bomb/.

Brown, John Seely, and Paul Duguid. *The Social Life of Information.* Boston: Harvard Business School Press, 2000.

Bruns, Axel. *Blogs, Wikipedia, Second Life, and Beyond: From Production to Produsage.* New York: Peter Lang, 2008.

Bryson, Bill. *Notes from a Small Island.* New York: Perennial, 2001.

Burgess, Jean, and Joshua Green. *YouTube: Online Video and Participatory Culture.* Malden, MA: Polity, 2009.

BuzzFeed. "About." BuzzFeed. Accessed 29 December 2012. http://www.buzzfeed.com/about.

Callon, Michel. "Some Elements of a Sociology of Translation: Domestication of the Scallops and the Fishermen of St Brieuc Bay." In *Power, Action and Belief: A New Sociology of Knowledge?* edited by John Law, 196–223. London: Routledge, 1986.

———. "Society in the Making: The Study of Technology as a Tool for Sociological Analysis." In *The Social Construction of Technological Systems,* edited by Weibe E. Bijker, Thomas P. Hughes, and Trevor E. Pinch, 85–103. Boston: MIT Press, 1987.

———. "Actor-Network Theory—The Market Test." In *Actor Network Theory and After,* edited by John Law and John Hassard, 181–95. Reprint, Oxford: Blackwell / *Sociological Review,* (1999) 2005.

Cantoni, Lorenzo, and Stefano Tardini. *Internet.* New York: Routledge, 2006.

Carey, John. "The Functions and Uses of Media during the September 11 Crisis and Its Aftermath." In *Crisis Communications: Lessons from September 11,* edited by A. Michael Noll, 1–16. Lanham, MD: Rowman and Littlefield, 2003.

Chief Investigating Officer, Government of India. In the Court of Addl. Ch. M. M., 37th Court, Esplanade, Mumbai. "Final Form / Report" (under Section 173 Cr. P. C.), 25 February 2009. [Online]. Available at: www.hindu.com/nic/mumbai-terror-attack-final-form.pdf.

CNN. "Hurricane Katrina Safe List." CNN, 2005. Accessed 7 February 2007 (retrieved archive), http://www.cnn.com/SPECIALS/2005/hurricanes/list/ (page discontinued April 2006).

———. "The Missing, the Found, the Victims." CNN. Last modified 28 January 2010. http://haiticrisis.cnn.com/.

"CNN-Dina Mehta.avi." YouTube video, 3:55, interview of Dina Mehta by Jim Clancy at CNN International, posted by dinamehta, 27 November 2008, http://youtu.be/LwAKSxO0KPI.

Collins, Randall. "Rituals of Solidarity and Security in the Wake of Terrorist Attack." *Sociological Theory* 22, no. 1 (2004): 53–87.

comScore. "comScore Releases August 2011 U.S. Online Video Rankings," news release, 22 September 2011, http://www.comscore.com/Press_Events/Press_Releases/2011/9/comScore_Releases_August_2011_U.S._Online_Video_Rankings.

Cooper, Alan. *About Face: The Essentials of User Interface Design.* Foster City, CA: IDG Books Worldwide, 1995.

Cooper, Alan, Robert Reimann, and David Cronin. *About Face 3: The Essentials of Interaction Design.* Indianapolis: Wiley, 2007.

Courage, Catherine, and Kathy Baxter. *Understanding Your Users: A Practical Guide to User Requirements; Methods, Tools, and Techniques.* San Francisco: Morgan Kaufmann, 2005.

Cridland, James. (2005). Emergency in London (Set). Flickr. Retrieved from http://www.flickr.com/photos/jamescridland/sets/554970/.

Cronin, James. (2005). 07072005.jpg. Flickr. Retrieved from http://www-us.flickr.com/photos/jamescronin/24231176/in/pool-bomb/.

Crump, Imogen. "Using Twitter to Cover the Earthquake in Japan." *BBC Radio 5 live* (blog), 14 March 2011, http://www.bbc.co.uk/blogs/5live/2011/03/using-twitter-to-cover-the-ear.shtml.

Davis, Marjorie T. "Shaping the Future of Our Profession." *Technical Communication* 48, no. 2 (2001): 139–44.

Dennen, Alfie. "London Underground Bombing, Trapped." *Alfies Moblog.* Last modified 7 July 2005, http://moblog.net/view/77571/london-underground-bombing-trapped.

DeWalt, Kathleen M., and Billie R. DeWalt. "Participant Observation." With Coral B. Wayland. In *Handbook of Methods in Cultural Anthropology,* edited by H. Russell Bernard, 259–99. Walnut Creek, CA: AltaMira, 1998.

Dodge, Martin. "The Role of Maps in Virtual Research Methods." In *Virtual Methods: Issues in Social Research on the Internet,* edited by Christine Hine, 113–128. New York: Berg, 2005.

Doheny-Farina, Stephen. *The Wired Neighborhood.* New Haven, CT: Yale University Press, 1996.

Donath, Judith S. "Identity and Deception in the Virtual Community." In *Communities in Cyberspace,* edited by Mark A. Smith and Peter Kollock, 29–59. London: Routledge, 1999.

Drexler, K. Eric. "Hypertext Publishing and the Evolution of Knowledge." *Social Intelligence* 1, no. 2 (1991): 87–120, http://e-drexler.com/d/06/00/Hypertext/HPEK0.html.

Dubinsky, James M. Introduction to *Teaching Technical Communication: Critical Issues for the Classroom,* edited by James M. Dubinsky, 1–10. Boston: Bedford/St. Martin's, 2004.

Ehn, Pelle. *Work-Oriented Design of Computer Artifacts.* Hillsdale, NJ: Lawrence Erlbaum, 1989.

El-Nawawy, Mohammed, and Sahar Khamis. "Political Activism 2.0: Comparing the Role of Social Media in Egypt's 'Facebook Revolution' and Iran's 'Twitter Uprising.'" *Cyber-Orient* 6, no. 1 (2012), http://www.cyberorient.net/article.do?articleId=7439.

Ervin, Alexander M. *Applied Anthropology: Tools and Perspectives for Contemporary Practice.* Boston: Allyn and Bacon, 2000.

Ess, Charles. "The Political Computer: Democracy, CMC, and Habermas." In *Philosophical Perspectives on Computer-Mediated Communication,* edited by Charles Ess, 197–230. Albany: State University of New York Press, 1996.

Facebook. "Key Facts." Facebook Newsroom. Accessed 20 November 2012, http://newsroom.fb.com/Key-Facts.

Farivar, Cyrus. "Social Media, Web Provide Japan Quake Resources." Deutsche Welle, 3 November 2011, http://www.dw.de/social-media-web-provide-japan-quake-resources/a-14905640.

FBI. "Special Agent in Charge Richard DesLauriers Speaks at Press Conference Regarding Boston Marathon Explosions." Press Release from the FBI, 16 April 2013. Retrieved from http://www.fbi.gov/boston/press-releases/2013/special-agent-in-charge-richard-deslauriers-speaks-at-press-conference-regarding-boston-marathon-explosions.

Fischer, Gerhard. "Understanding, Fostering, and Supporting Cultures of Participation." *Interactions* 18, no. 3 (2011): 42–53, http://l3d.cs.colorado.edu/~gerhard/papers/2011/interactions-coverstory.pdf.

Flickr. "London Bomb Blasts Community / Pool / Tags." Flickr. Accessed 2008, www.flickr.com/groups/bomb/pool/tags/.

Forsythe, Diana E. "It's Just a Matter of Common Sense: Ethnography as Invisible Work." *Computer Supported Cooperative Work* 8 (1999): 127–45.

Fowler, Martin. *UML Distilled: A Brief Guide to the Standard Object Modeling Language.* 3rd ed. Reading, MA: Addison-Wesley, 2003.

Francis, Mark Norman. (2005). Charing Cross Road is cordoned off. Flickr. Retrieved from http://www-us.flickr.com/photos/mn_francis/24227078/in/pool-bomb/.

Friedman, Thomas L. *The World Is Flat: A Brief History of the Twenty-First Century.* New York: Farrar, Straus and Giroux, 2005.

Fuller, Jacquelline, and Prem Ramaswami (for the Google Crisis Response Team). "Staying Connected in Post-Earthquake Haiti." *Google Official Blog,* 15 January 2010, http://googleblog.blogspot.com/2010/01/staying-connected-in-post-earthquake.html.

Garrett, Jesse James. *The Elements of User Experience: User-Centered Design for the Web.* Indianapolis: New Riders, 2003.

Gauntlett, David. *Web Studies: Rewiring Media Studies for the Digital Age.* London: Arnold, 2000.

Geisler, Cheryl. "IText Revisited: The Continuing Interaction of Information Technology and Text." *Journal of Business and Technical Communication* 25, no. 3 (2011): 251–55.

GeoNet. "M 7.1, Darfield (Canterbury), 4 September 2010." GeoNet. Accessed 30 December 2012, http://info.geonet.org.nz/display/quake/M+7.1%2C+Darfield+%28Canterbury%29%2C+4+September+2010.

Google Person Finder. Accessed 30 December 2012, http://google.org/personfinder/global/home.html.

Grabill, Jeffrey T. "Knowledge Work and Technical Communication as Public Rhetoric," presented at the Conference of the Association of Teachers of Technical Writing, San Francisco, CA, 2009.

Grabill, Jeffrey T., and W. Michele Simmons. "Toward a Critical Rhetoric of Risk Communication: Producing Citizens and the Role of Technical Communication." *Technical Communication Quarterly* 7, no. 4 (1998): 415–41.

Hackos, JoAnn T., and Redish, Janice C. *User and Task Analysis for Interface Design.* New York: Wiley, 1998.

Haiticrisis.appspot.com. Accessed 29 December 2012, http://www.haiticrisis.appspot. com/ (site discontinued).

Halavais, Alexander. *Search Engine Society.* Malden, MA: Polity, 2009.

Halloran, S. Michael. "Aristotle's Concept of Ethos, or If Not His Somebody Else's." *Rhetoric Review* 1, no. 1 (1982): 58–63. doi: 10.1080/07350198209359037.

Halvorson, Kristina, and Melissa Rach. *Content Strategy for the Web.* 2nd ed. Berkeley, CA: New Riders, 2012.

Hart-Davidson, William. "On Writing, Technical Communication, and Information Technology: The Core Competencies of Technical Communication." *Technical Communication* 48 (2001): 145–55.

Herring, Susan C. "Gender and Democracy in Computer-Mediated Communication." *Electronic Journal of Communication* 3, no. 2 (1993), Retrieved from http://www.cios. org/EJCPUBLIC/003/2/00328.HTML.

Herrman, John. "Twitter Is a Truth Machine." BuzzFeed, accessed 30 December 2012, http://www.buzzfeed.com/jwherrman/twitter-is-a-truth-machine.

Hine, Christine. "Virtual Methods and the Sociology of Cyber-Social-Scientific Knowledge". In *Virtual Methods: Issues in Social Research on the Internet,* edited by Christine Hine, 1–13. New York: Berg, 2005.

———. "How Can Qualitative Internet Researchers Define the Boundaries of Their Projects?" In *Internet Inquiry: Conversations about Method,* edited by Annette N. Markham and Nancy K. Baym, 1–20. Thousand Oaks, CA: Sage, 2009.

The Home Office. *Report of the Official Account of the Bombings in London on 7th July 2005.* London: The Stationery Office, 11 May 2006, http://www.official-documents.gov.uk/ document/hc0506/hc10/1087/1087.pdf.

Howard, John. (2005). Evacuation. Flickr. Retrieved from www.flickr.com/photos/ jmh/24223193/.

Ingram, Mathew. 2012a. "Hurricane Sandy and Twitter as a Self-Cleaning Oven for News." GigaOM, 30 October, http://gigaom.com/2012/10/30/hurricane-sandy-and-twitter-as-a-self-cleaning-oven-for-news/.

———. 2012b. "It's Not Twitter—This Is Just the Way News Works Now." GigaOM, 15 December, http://gigaom.com/2012/12/15/its-not-twitter-this-is-just-the-way-the-news-works-now/.

Intelligence and Security Committee. *Report into the London Terrorist Attacks on 7 July 2005,* Presented to Parliament by the Prime Minister by Command of Her Majesty, 11 May 2006, accessed at http://www.official-documents.gov.uk/document/cm67/6785/ 6785.pdf.

International Committee of the Red Cross. "Haiti 2010 Earthquake I Am Alive." International Committee of the Red Cross. Accessed 8 February 2010, http://www. icrc.org/web/doc/siterfl0.nsf/htmlall/familylinks-haiti-eng?opendocument (page discontinued).

Ito, Joi. "London Explosions." *Joi Ito* (blog), 7 July 2005, http://joi.ito.com/archives/2005/ 07/07/london_explosions.html.

Jenkins, Henry. *Convergence Culture: Where Old and New Media Collide.* New York: New York University Press, 2006.

————. *Confronting the Challenges of Participatory Culture: Media Education for the 21st Century*. With Ravi Purushotma, Margaret Weigel, Katie Clinton, and Alice J. Robison. Cambridge, MA: MIT Press, 2009. http://mitpress.mit.edu/sites/default/files/titles/free_download/9780262513623_Confronting_the_Challenges.pdf.

Johnson, Robert R. "Audience Involved: Toward a Participatory Model of Writing." *Computers and Composition* 14, no. 3 (1997): 361–76.

————. *User-Centered Technology: A Rhetorical Theory for Computers and Other Mundane Artifacts*. Albany: State University of New York Press, 1998.

Johnson-Eilola, Johndan. *Datacloud: Toward a New Theory of Online Work*. Cresskill, NJ: Hampton Press, 2005.

Jones, Calvert, and Sarai Mitnick. "Open Source Disaster Recovery: Case Studies of Networking Collaboration." *First Monday* 11, no. 5 (2006), http://www.firstmonday.org/htbin/cgiwrap/bin/ojs/index.php/fm/article/view/1325/1245.

Jones, Steve, ed. *Doing Internet Research: Critical Issues and Methods for Examining the Net*. Thousand Oaks, CA: Sage, 1999.

Kimball, Miles A., and Ann R. Hawkins. *Document Design: A Guide for Technical Communicators*. Boston: Bedford/St. Martin's, 2004.

Knabb, Richard D., Jamie R. Rhome, and Daniel P. Brown. *Tropical Cyclone Report: Hurricane Katrina; 23–30 August 2005*. National Hurricane Center. 20 December 2005, http://www.nhc.noaa.gov/pdf/TCR-AL122005_Katrina.pdf.

Kock, Ned. *Systems Analysis and Design Fundamentals: A Business Process Redesign Approach*. Thousand Oaks, CA: Sage, 2007.

Kotz, Deborah. "Injury Toll from Marathon Bombings Rises." *Boston Globe*, 23 April 2013. http://www.bostonglobe.com/metro/massachusetts/2013/04/22/just-bombing-victims-still-critically-ill-but-count-injured-rises/7mUGAu5tJgKsxc634NCAJJ/story.html.

krisnair. (2008). Twitter / @Fenriq RT @narendranag: #mumbai. if you need help . . . Twitter. Retrieved 26 November 2008, from http://twitter.com/#!/krisnair/statuses/ (tweet no longer available).

Krug, Steve. *Don't Make Me Think! A Common Sense Approach to Web Usability*. 2nd ed. Berkeley, CA: New Riders, 2006.

Latour, Bruno. *Science in Action: How to Follow Scientists and Engineers through Society*. Reprint, Cambridge, MA: Harvard University Press, (1987) 2003.

————. *Aramis, or the Love of Technology*. Translated by Catherine Porter. Reprint, Cambridge, MA: Harvard University Press, (1996) 2002.

————. (1999a) 2005. "On Recalling ANT." In *Actor Network Theory and After*, edited by John Law and John Hassard, 15–25. Reprint, Oxford: Blackwell / Sociological Review.

————. 1999b. *Pandora's Hope: Essays on the Reality of Science Studies*. Cambridge, MA: Harvard University Press.

————. *Reassembling the Social: An Introduction to Actor-Network-Theory*. New York: Oxford University Press, 2005.

Laurel, Brenda. "Muscular Design." Introduction to *Design Research: Methods and Perspectives*, edited by Brenda Laurel, 16–19. Cambridge, MA: MIT Press, 2003.

Law, John. "Notes on the Theory of the Actor-Network: Ordering, Strategy and Heterogeneity." *Systems Practice* 5 (1992): 379–93.

————. *After Method: Mess in Social Science Research*. New York: Routledge, 2004.

Law, John, and Annemarie Mol. *Situating Technoscience: An Inquiry into Spatialities*. Lancaster LA1 4YN, UK: Centre for Science Studies, Lancaster University. Last modified 8 December 2003, http://www.lancs.ac.uk/fass/sociology/papers/law-mol-situating-technoscience.pdf.

Lawley, Liz. "Blog Research Issues." *Many2Many* (Corante blog), 24 June 2004, http://many.corante.com/archives/2004/06/24/blog_research_issues.php.

Leung, Denise. "Say Hello to the New Flickr Uploadr." *Flickr Blog*, 25 April 2012, http://blog.flickr.net/en/2012/04/25/say-hello-to-the-new-flickr-uploadr/.

Lévy, Pierre. *Collective Intelligence: Mankind's Emerging World in Cyberspace.* Translated by Robert Bononno. Cambridge, MA: Helix Books / Perseus Books, 1997.

London Bloggers. "Rachel from North London." London Bloggers. 27 February 2006. Retrieved from http://londonbloggers.iamcal.com/weblog.php?id=2715.

London Bomb Blasts Community. "London Bomb Blasts Community" (Flickr group). Flickr. Accessed 2008, http://www.flickr.com/groups/bomb/.

Martinez, Michael. "Joplin Twister's Death Toll Rises to 142," image collection. *CNN*, 28 May 2011, http://www.cnn.com/2011/US/05/28/missouri.tornado/index.html.

Martinovic, Emma. "Norway Attacks: A Survivor's Account of the Utøya Massacre." Translated by Andrew Boyle. *Guardian*, 27 July 2011, http://www.guardian.co.uk/world/2011/jul/27/norway-attacks-survivor-utoya-massacre.

McNely, Brian. "Backchannel Persistence and Collaborative Meaning-Making," in *Proceedings of the 27th ACM International Conference on Design of Communication*, 2009, 297–303.

Mehta, Dina. (2008a). "Commentary: How Social Media Shared Pain and Rage of Mumbai." *CNN*, last modified December 2, http://www.cnn.com/2008/WORLD/asiapcf/12/02/mehta.mumbai/index.html.

———. (2008b). "Terror Victims—27th Nov 2008" (Google spreadsheet). Google Docs. http://spreadsheets.google.com/pub?key=p_esnE-3Z3p-HehX1YOZIaw.

———. (2008c). Twitter / @Dina Mehta: #mumbai—help needed . . . Twitter. Retrieved from http://twitter.com/dina/statuses/1026464551.

Miller, Carolyn R. "A Humanistic Rationale for Technical Writing." *College English* 40, no. 6 (1979): 610–17.

Mirel, Barbara. "Writing and Database Technology: Extending the Definition of Writing in the Workplace." In *Electronic Literacies in the Workplace: Technologies of Writing,* edited by Patricia Sullivan and Jennie Dautermann, 91–114. Urbana, IL: National Council of Teachers of English and Computers and Composition, 1996.

Moggan, Stephan. (2005). Helicopter. Flickr. Retrieved from http://www-us.flickr.com/photos/moggan/24233559/in/pool-bomb/.

Mol, Annemarie, and John Law. "Regions, Networks and Fluids: Anaemia and Social Topology." *Social Studies of Science* 24 (1994): 641–71.

morealtitude. (2010). Twitter / @louis_press: Note ur tweeting . . . Twitter. Retrieved from https://twitter.com/morealtitude (tweet no longer available).

Morville, Peter. (2004). "Three Circles of Information Architecture" (image). *Semantic Studios*. Retrieved from http://semanticstudios.com/publications/semantics/images/threecirclesbig.jpg.

———. *Ambient Findability: What We Find Changes Who We Become.* Sebastopol, CA: O'Reilly Media, 2005.

Morville, Peter, and Rosenfeld, Louis. *Information Architecture for the World Wide Web: Designing Large-Scale Web Sites.* 3rd ed. Sebastopol, CA: O'Reilly Media, 2006.

Mumbai Help. November 2008 Archives. *Mumbai Help* (blog), November 2008, http://mumbaihelp.blogspot.com/2008_11_01_archive.html.

New York Times. 2012a. "Haiti." *New York Times,* last modified 24 December, http://topics.nytimes.com/top/news/international/countriesandterritories/haiti/index.html.

————. 2012b. "Hurricane Sandy: Covering the Storm." *New York Times,* last modified 6 November, http://www.nytimes.com/interactive/2012/10/28/nyregion/hurricane-sandy.html.

NGDC (National Geophysical Data Center). "Great Tohoku, Japan Earthquake and Tsunami, 11 March 2011." NOAA National Geophysical Data Center. Accessed on 30 December 2012, http://www.ngdc.noaa.gov/hazard/honshu_11mar2011.shtml.

NOAA (National Oceanic and Atmospheric Administration). "NOAA Reviews Record-Setting 2005 Atlantic Hurricane Season: Active Hurricane Era Likely to Continue." National Oceanic and Atmospheric Administration, 29 November 2005, http://www.noaanews.noaa.gov/stories2005/s2540.htm.

Norman, Donald A. *The Design of Everyday Things.* New York: Doubleday, 1989.

Olson, Katie. (2008). "I'm ok at VT" (Facebook group). Facebook. Retrieved from http://mbc.facebook.com/group.php?gid = 2321223134 (page discontinued).

oops777. Find Boston Bombers. Reddit. Accessed 2013. Retrieved from http://www.reddit.com/r/findbostonbombers (site made private).

Perlmutter, David D. "Katrina: Too Close to Home." *Critical Studies in Media Communication* 23 (2006): 78–80.

Pew Internet and American Life Project. 2012a. "Trend Data (Adults): Device Ownership." Pew Internet and American Life Project. Accessed 30 December, http://pewinternet.org/Trend-Data-%28Adults%29/Device-Ownership.aspx.

————. 2012b. "Trend Data (Adults): Who's Online; Internet User Demographics." Pew Internet and American Life Project. Accessed 30 December, http://pewinternet.org/Static-Pages/Trend-Data-(Adults)/Whos-Online.aspx.

Potts, Liza. "Diagramming with Actor Network Theory: A Method for Modeling Holistic Experience," in *Proceedings of the International Professional Communication Conference,* 2008, 1–6.

————. 2009a. "Designing for Disaster: Social Software Use in Times of Crisis." *International Journal of Sociotechnology and Knowledge Development* 1, no. 2: 33–46.

————. 2009b. "Experience Design for Participation," in *Proceedings of the Sociotech Workshop at Interact '09 Conference,* 1–5.

————. 2009c. "Peering into Disaster: Social Software Use from the Indian Ocean Earthquake to the Mumbai Bombings," in *Proceedings of the International Professional Communication Conference,* 1–8.

————. 2009d. "Using Actor Network Theory to Trace and Improve Multimodal Communication Design." *Technical Communication Quarterly* 18, no. 3: 281–301.

————. "Consuming Digital Rights: Mapping the Artifacts of Entertainment." *Technical Communication* 57, no. 3 (2010): 300–18.

Potts, Liza, and Gerianne Bartocci. "<Methods>Experience Design</Methods>," in *Proceedings of the 27th ACM International Conference on Design of Communication,* 2009, 17–22.

Potts, Liza, and Dave Jones. "Contextualizing Experiences: Tracing the Relationships between People and Technologies in the Social Web." *Journal of Business and Technical Communication* 25, no. 2 (2011): 1–21.

Potts, Liza, Joyce Seitzinger, Dave Jones, and Angela Harrison. "Tweeting Disaster: Hashtags Constructions and Collisions," in *Proceedings of the 29th ACM International Conference on Design of Communication,* 2011, 235–40.

Powazek, Derek M. "What the Hell is a Weblog and Why Won't They Leave Me Alone? A Personal Opinion by Derek M. Powazek." Powazek, 17 February 2000, http://powazek.com/wtf/.

pradx. (2008a). Twitter / @Pradeep Mohandas: #mumbai. See if . . . Twitter. Retrieved from http://twitter.com/#!/pradx/statuses/1025211644.

———. (2008b). Twitter / @Pradeep Mohandas: Important contact nos of . . . Twitter. Retrieved from http://twitter.com/#!/pradx/statuses/1025194693.

Press Information Bureau, Government of India. "HM Announces Measures to Enhance Security," news release, 11 December 2008, http://pib.nic.in/release/release.asp?relid=45446.

Putnam, Laurie. "By Choice or by Chance: How the Internet Is Used to Prepare for, Manage, and Share Information about Emergencies." *First Monday* 7, no. 11 (2002), http://firstmonday.org/htbin/cgiwrap/bin/ojs/index.php/fm/article/view/1007/928.

Rainie, Lee, and Barry Wellman. *Networked: The New Social Operating System.* Cambridge, MA: MIT Press, 2012.

Rand Corporation. "Paul Baran and the Origins of the Internet." Rand Corporation. Last modified 23 December 2011. http://www.rand.org/about/history/baran.html.

Rappoport, Paul N., and James Alleman. "The Internet and the Demand for News: Macro- and Microevidence." In *Crisis Communications: Lessons from September 11,* edited by A. Michael Noll, 149–66. Lanham, MD: Rowman and Littlefield, 2003.

Ratcliffe, Lindsay, and Marc McNeill. *Agile Experience Design: A Digital Designer's Guide to Agile, Lean, and Continuous.* Berkeley, CA: New Riders, 2012.

reddit. "Reflections on the Recent Boston Crisis." *reddit Blog,* 22 April 2013. Retrieved from http://blog.reddit.com/2013/04/reflections-on-recent-boston-crisis.html.

Redish, Janice (Ginny). *Letting Go of the Words: Writing Web Content That Works.* Amsterdam: Elsevier / Morgan Kaufmann, 2007.

Rheingold, Howard. *The Virtual Community: Homesteading on the Electronic Frontier.* Reading, MA: Addison-Wesley, 1993.

Roberto, Frankie. (2005). BBC Homepage Timeout. Flickr. Retrieved from http://www.flickr.com/photos/frankieroberto/24229068.

rockmother. (2005). Emergency number. Flickr. Retrieved from http://www.flickr.com/photos/85634220@N00/24250353/in/pool-bomb/.

Roos, John V. (2011). Twitter / @AmbassadorRoos: If you have friends . . . Twitter. Retrieved from https://twitter.com/AmbassadorRoos/status/46222131775799296.

Rutter, Jason, and Gregory W. H. Smith. "Ethnographic Presence in a Nebulous Setting." In *Virtual Methods: Issues in Social Research on the Internet,* edited by Christine Hine, 81–92. New York: Berg, 2005.

Saffer, Dan. *Designing for Interaction: Creating Innovative Applications and Devices.* Berkeley, CA: New Riders, 2009.

Savage, Gerald J. "Tricksters, Fools, and Sophists: Technical Communication as Postmodern Rhetoric." In *Power and Legitimacy in Technical Communication Volume II: Strategies for Professional Status,* edited by Teresa Kynell-Hunt and Gerald J. Savage, 167–93. Amityville, NY: Baywood, 2004.

scorpfromhell. (2008). Twitter / @A. Prem Kumar: car no . . . Twitter. Retrieved from http://twitter.com/#!/scorpfromhell/statuses/1025741477.

Seitzinger, Joyce. "Social Media Use in Crisis—#eqnz—Which Hashtag Prevails?" *Cat's Pyjamas* (blog), 4 September 2010, http://www.cats-pyjamas.net/2010/09/social-media-use-in-a-crisis-eqnz-which-hashtag-prevails/.

Slattery, Shaun. "Undistributing Work through Writing: How Technical Writers Manage Texts in Complex Information Environments." *Technical Communication Quarterly* 16, no. 3 (2007): 311–25.

Sollid, Stephen J. M., Rune Rimstad, Marius Rehn, Anders R. Nakstad, Ann-Elin Tomlin-son, Terje Strand, Hans J. Heimdal, Jan E. Nilsen, and Mårten Sandberg, and collabo-rating group. "Oslo Government District Bombing and Utøya Island Shooting July 22, 2011: The Immediate Prehospital Emergency Medical Service Response." *Scandina-vian Journal of Trauma, Resuscitation and Emergency Medicine* 20, no. 3 (2012), http://www.sjtrem.com/content/20/1/3.

Spinuzzi, Clay. "Toward Integrating Our Research Scope: A Sociocultural Field Methodol-ogy." *Journal of Business and Technical Communication* 16 (2002): 3–23.

———. *Tracing Genres through Organizations: A Sociocultural Approach to Information Design.* Cambridge, MA: MIT Press, 2003.

———. "The Methodology of Participatory Design." *Technical Communication* 52, no. 2 (2005): 163–74.

———. "Guest Editor's Introduction: Technical Communication in the Age of Distrib-uted Work." *Technical Communication Quarterly* 16, no. 3 (2007): 265–77.

———. *Network: Theorizing Knowledge Work in Telecommunications.* Cambridge: Cam-bridge University Press, 2008.

———. "Symmetry as a Methodological Move, Part 1." *Spinuzzi: A Blog about Rhetoric, Technology, Research, and Where We're Headed Next* (blog), 18 March 2012a, http://spinuzzi.blogspot.com/2012/03/symmetry-as-methodological-move-part-i.html.

———. "Symmetry as a Methodological Move, Part 2." *Spinuzzi: A Blog about Rhetoric, Technology, Research, and Where We're Headed Next* (blog), 20 March 2012b, http://spinuzzi.blogspot.com/2012/03/symmetry-as-methodological-move-part-ii.html.

Spolsky, Joel. "It's Not Just Usability." *Joel on Software,* September 6, 2004, http://www.joelonsoftware.com/articles/NotJustUsability.html.

SSSIndore. (2008). Twitter / @SaptagiriS: #bomb #bombay #mumbai . . . Twitter. Retrieved from http://twitter.com/#!/shrik_shrek/statuses/1026290847.

Stocking, George W., Jr. "History of Anthropology: Whence/Whither." In *Observers Observed: Essays on Ethnographic Fieldwork,* edited by George W. Stocking Jr., 3–12. Madison: University of Wisconsin Press, 1983.

Stolley, Karl. "Integrating Social Media into Existing Work Environments." *Journal of Busi-ness and Technical Communication* 23, no. 3 (2009): 350–71.

Storey, David. (2005a). "About fgt" (user profile page). *Flickr.* Retrieved 2005, http://www-us.flickr.com/people/fgt/ (site discontinued).

———. (2005b). Flickr comment on "kingscross" (photo), uploaded on 7 July by Antarc-tic Lemur. *Flickr.* Retrieved from http://www.flickr.com/photos/antarcticlemur/24228026/#comment6701975.

Stott, Pete. (2005a). Police everywhere. *Flickr.* Retrieved from http://www.flickr.com/photos/snowcrash/24244802/in/pool-bomb/.

———. (2005b). Special forces / river police. *Flickr.* Retrieved from http://www.flickr.com/photos/snowcrash/24245088/in/pool-bomb/.

Sullivan, Dale L. "Political-Ethical Implications of Defining Technical Communication as a Practice." *Journal of Advanced Composition* 10 (1990): 375–86.

Swarts, Jason. "Mobility and Composition: The Architecture of Coherence in Non-Places." *Technical Communication Quarterly* 16, no. 3 (2007): 279–309.

———. *Together with Technology: Writing Review, Enculturation, and Technological Medi-ation.* Amityville, NY: Baywood, 2008.

———. "The Collaborative Construction of 'Fact' on Wikipedia," in *Proceedings of the 27th ACM International Conference on Design of Communication,* 2009, 281–88.

———. "Recycled Writing: Assembling Actor-Networks from Reusable Content." *Journal of Business and Technical Communication* 24, no. 2 (2010): 127–63.

"Talk: Sandy Hook Elementary School Shooting/Archive 3: Victims Section." *Wikipedia*. Accessed 30 December 2012. http://en.wikipedia.org/wiki/Talk:Sandy_Hook_Elementary_School_shooting/Archive_3#Victims_section.

Taylor, Chris. "Twitter Users React to Massive Quake, Tsunami in Japan." *Mashable*, 10 March 2011, http://mashable.com/2011/03/11/japan-tsunami/.

Tebeaux, Elizabeth, and Sam Dragga. *The Essentials of Technical Communication*. New York: Oxford University Press, 2010.

Thompson, Matt. "How the Virginia Tech Shooting Changed The Washington Post's Reporting and Online Publishing." *Poynter*, 23 June 2010, http://www.poynter.org/uncategorized/103954/how-the-virginia-tech-shooting-changed-the-washington-posts-reporting-and-online-publishing/.

Tinworth, Adam. (2005). Police on the move. Flickr. Retrieved from http://www-us.flickr.com/photos/adders/24230030/in/pool-bomb/.

Torkington, Nat. "Homophily in Social Software." *O'Reilly Radar*, 17 October 2006, http://radar.oreilly.com/2006/10/homophily-in-social-software.html.

Transport for London. Accessed 30 December 2012. http://www.tfl.gov.uk/.

tristamsparks. (2010). Twitter / thanks megan! RT @harvestbird Try tag #eqnz … Twitter. Retrieved 2010 from https://twitter.com/tristamsparks (tweet no longer available).

Turkle, Sherry. *Life on the Screen: Identity in the Age of the Internet*. New York: Simon and Schuster, 1995.

Twitter for Business. "What Is Twitter?" Twitter for Business. Accessed 24 December 2012. https://business.twitter.com/basics/what-is-twitter/.

United Nations Office of the Special Envoy for Tsunami Recovery. "The Human Toll." Accessed 2010. http://www.tsunamispecialenvoy.org/country/humantoll.asp.

Vander Wal, Thomas. "Social Software Design for One" (PowerPoint presentation, DCampSouth, Raleigh, NC, 2 June 2007). Retrieved from http://www.slideshare.net/vanderwal/math-of-social-software-in-designing-social-software-for-one.

van Dijck, José. "Users Like You? Theorizing Agency in User-Generated Content." *Media, Culture & Society* 31, no. 1 (2009): 41–58.

Vargas, Jose Antonio. "A Chain of Grief with Links on Facebook." *Washington Post*, 18 April 2007, http://www.washingtonpost.com/wp-dyn/content/article/2007/04/17/AR2007041702037.html.

Vengerfeldt, Pille. "The Internet as a News Medium for the Crisis News of Terrorist Attacks in the United States." In *Crisis Communications: Lessons from September 11*, edited by A. Michael Noll, 133–48. Lanham, MD: Rowman and Littlefield, 2003.

Virginia Tech Review Panel. *Mass Shootings at Virginia Tech April 16, 2007* (report presented to Timothy M. Kaine, Governor Commonwealth of Virginia), August 2007, http://www.governor.virginia.gov/TempContent/techPanelReport-docs/FullReport.pdf.

Weinberger, David. *Too Big to Know: Rethinking Knowledge Now That the Facts Aren't the Facts, Experts Are Everywhere, and the Smartest Person in the Room Is the Room*. New York: Basic Books, 2012.

Wellman, Barry. "An Electronic Group Is Virtually a Social Network." In *Culture of the Internet*, edited by Sara Kiesler, 179–205. Mahwah, NJ: Lawrence Erlbaum, 1997.

Wikipedia. "Statistics." *Wikipedia*. Accessed 20 November 2012. http://en.wikipedia.org/wiki/Special:Statistics.

Winn, Patrick. "Japan Tsunami Disaster: As Japan Scrambles, Twitter Reigns." *GlobalPost*, 18 March 2011, http://www.globalpost.com/dispatch/news/regions/asia-pacific/japan/110318/twitter-japan-tsunami.

Wortham, Jenna. "After 10 Years of Blogs, the Future's Brighter Than Ever." *Wired*, 17 December 2007. http://www.wired.com/entertainment/theweb/news/2007/12/blog_anniversary.

Zappen, James P. *The Rebirth of Dialogue: Bakhtin, Socrates, and the Rhetorical Tradition.* Albany: State University of New York Press, 2004.

Zappen, James P., and Cheryl Geisler. "Designing the Total User Experience: Implications for Research and Program Development." *Programmatic Perspectives* 1, no. 1 (2009): 3–28.

zigzackly. "Can We Help?" *Mumbai Help* (blog), 27 November 2008, http://mumbaihelp. blogspot.com/2008/11/can-we-help.html.

INDEX

For Product Safety Concerns and Information please contact our EU
representative GPSR@taylorandfrancis.com
Taylor & Francis Verlag GmbH, Kaufingerstraße 24, 80331 München, Germany

www.ingramcontent.com/pod-product-compliance
Lightning Source LLC
Chambersburg PA
CBHW071000050326
40689CB00014B/3430